The Fundamentals of Analysis for Talented Freshmen

Synthesis Lectures on Mathematics & Statistics

Editor
Steven G. Krantz, *Washington University, St. Louis*

The Fundamentals of Analysis for Talented Freshmen
Peter M. Luthy, Guido L. Weiss, and Steven S. Xiao
2016

Aspects of Differential Geometry II
Peter Gilkey, JeongHyeong Park, and Ramón Vxiázquez-Lorenzo
2015

Aspects of Differential Geometry I
Peter Gilkey, JeongHyeong Park, and Ramón Vázquez-Lorenzo
2015

An Easy Path to Convex Analysis and Applications
Boris S. Mordukhovich and Nguyen Mau Nam
2013

Applications of Affine and Weyl Geometry
Eduardo García-Río, Peter Gilkey, Stana Nikcevic, and Ramón Vázquez-Lorenzo
2013

Essentials of Applied Mathematics for Engineers and Scientists, Second Edition
Robert G. Watts
2012

Chaotic Maps: Dynamics, Fractals, and Rapid Fluctuations
Goong Chen and Yu Huang
2011

Matrices in Engineering Problems
Marvin J. Tobias
2011

The Integral: A Crux for Analysis
Steven G. Krantz
2011

Statistics is Easy! Second Edition
Dennis Shasha and Manda Wilson
2010

Lectures on Financial Mathematics: Discrete Asset Pricing
Greg Anderson and Alec N. Kercheval
2010

Jordan Canonical Form: Theory and Practice
Steven H. Weintraub
2009

The Geometry of Walker Manifolds
Miguel Brozos-Vázquez, Eduardo García-Río, Peter Gilkey, Stana Nikcevic, and Ramón Vázquez-Lorenzo
2009

An Introduction to Multivariable Mathematics
Leon Simon
2008

Jordan Canonical Form: Application to Differential Equations
Steven H. Weintraub
2008

Statistics is Easy!
Dennis Shasha and Manda Wilson
2008

A Gyrovector Space Approach to Hyperbolic Geometry
Abraham Albert Ungar
2008

The Fundamentals of Analysis for Talented Freshmen

Peter M. Luthy, Guido L. Weiss, and Steven S. Xiao

ISBN: 978-3-031-01281-5   paperback
ISBN: 978-3-031-02409-2   ebook

DOI 10.1007/978-3-031-02409-2

A Publication in the Springer series
SYNTHESIS LECTURES ON MATHEMATICS & STATISTICS

Lecture #17
Series Editor: Steven G. Krantz, *Washington University, St. Louis*
Series ISSN
Print 1938-1743   Electronic 1938-1751

The Fundamentals of Analysis for Talented Freshmen

Peter M. Luthy
College of Mount Saint Vincent

Guido L. Weiss
Washington University in St. Louis

Steven S. Xiao
Washington University in St. Louis

SYNTHESIS LECTURES ON MATHEMATICS & STATISTICS #17

ABSTRACT

This book assumes the students know some of the basic facts about Calculus. We are very rigorous and expose them to the proofs and the ideas which produce them. In three chapters, this book covers these number systems and the material usually found in a junior-senior advanced Calculus course. It is designed to be a one-semester course for "talented" freshmen. Moreover, it presents a way of thinking about mathematics that will make it much easier to learn more of this subject and be a good preparation for more of the undergraduate curriculum.

KEYWORDS

cardinality, derivative, differentiable, gradient, limit, norm, partial derivative, power set, Riemann sum, uniformaly continuous, upper integral

Contents

List of Figures

Preface

The title of this book is *The Fundamentals of Analysis for Talented Freshmen*. It is based on a set of notes for a course developed over many years and taught by Guido Weiss to a class of talented freshmen at Washington University in St. Louis. This course was attended each fall semester by approximately twenty incoming students, most of whom became mathematics majors, and, subsequently, mathematicians. The course was designed for students with strong mathematical ability to develop their understanding of more advanced mathematics.

The three authors collaborated in the transformation of the course notes into book form. The book has three chapters. The first chapter was written by Guido Weiss and Peter Luthy. The second and third chapters were written by Guido Weiss and Steven Xiao with input from Peter Luthy.

We have assumed that students using this book already have been exposed to Calculus, and we will take advantage of this knowledge.

It is not our purpose to present these ideas from "first principles" and "the axioms of mathematics". We will, however, be very rigorous and introduce you to the proofs and the ideas which produce them. We will also provide considerable motivation for the topics we present to you.

Let us say a few words about some ideas which are familiar to you and will be important to understand calculus in one or more dimensions. You know several facts about the real numbers, $\mathbb{R}$ and the complex numbers, $\mathbb{C}$. They both contain the natural numbers, $\mathbb{N} = \{1, 2, 3, \cdots\}$, the integers $\mathbb{Z} = \{\cdots, -2, -1, 0, 1, 2, \cdots\}$, and the rational numbers, $\mathbb{Q} = \{\frac{m}{n} : m, n \in \mathbb{Z}, n \neq 0\}$. You know that two fractions $\frac{m}{n}$ and $\frac{p}{q}$, with $n, q \neq 0$, are equal if and only if $qm = pn$. You also are familiar with sets and a fair amount about set theoretic operations: the *union*, $S \cup T$ and the *intersection* $S \cap T$ of two sets S and T.

A function, $f : S \to T$, is defined to be the subset of the cartesian product $S \times T = \{(s, t) : s \in S, t \in T\}$ such that $(s, t) \in f$ if and only if $f(s) = t$ and, moreover, for every $s \in S$, there exists a unique $t \in T$ so that $t = f(s)$. For a function $f : S \to T$, the set S is called the *domain* of f and the *range* is the collection of all $t \in T$ satisfying the property that $f(s) = t$ for some s (the range of f is contained in T but might be a subset of T).

We will use many mathematical notations and list some of them below. We expect the student to be completely familiar with these symbols and their meanings.

	Symbol	Meaning
(1)	$\in$	an element of: if S is a set, then $x \in S$ means that x is an element of this set
(2)	$\subset$	contained in: if S and T are sets, then $S \subset T$ means that each element of S is an element of T
(3)	$\exists$	there exists
(4)	$\exists!$	there exists a unique
(5)	iff	if and only if
(6)	$p \Rightarrow q$	p implies q
(7)	$p \Leftarrow q$	q implies p
(8)	$p \Leftrightarrow q$	p if and only if q
(9)	$\ni$	such that
(10)	$\mathbb{N}$	the natural numbers: $\{1, 2, 3, \cdots\}$.
(11)	$\mathbb{Z}$	the integers: $\{\cdots, -3, -2, -1, 0, 1, 2, 3, \cdots\}$.
(12)	$\mathbb{Q}$	the rational numbers: the numbers represented by the fractions $\dfrac{m}{n}$, where m and n are integers, and n is nonzero
(13)	$\mathbb{R}$	the real numbers; clearly $\mathbb{N} \subset \mathbb{Z} \subset \mathbb{Q} \subset \mathbb{R}$
(14)	$f : S \to T$	a function whose domain is S with range $f(S) \subset T$; f can be considered as a subset of $S \times T$
(15)	$f : S \to T$ is injective (or 1:1 or one-to-one)	$f(s) = f(s')$ implies $s = s'$
(16)	$f : S \to T$ is surjective (or onto)	for each $t \in T$, there exists an $s \in S$ so that $f(s) = t$; in other words, $f(S) = T$
(17)	$f : S \to T$ is bijective or a one-to-one correspondence	f is both injective and surjective
(18)	$\blacksquare$	the proof is finished
(19)	$\varnothing$	the empty set

(20)	LUB	least upper bound
(21)	GLB	greatest lower bound
(22)	:=	defined to be equal to: for example $f'(x) := \lim\limits_{h \to 0} \dfrac{f(x + h) - f(x)}{h}$

(HW) indicates "Home Work."

The most important advice we can give you is this: read these pages carefully and do not lag behind the pace of this course. If you work hard and follow our advice, at the end of this course you will have covered much of the material presented in a junior-level course in analysis. Moreover, you will be exposed to a way of thinking about mathematics that will make it much easier for you to learn more of this subject.

We have divided the material contained in this book into three chapters. This introduction (or Preface), these three chapters, and an index not only represent the material of Calculus on $\mathbb{R}$ and $\mathbb{R}^n$ but also serve as good preparation for the rest of the undergraduate curriculum.

Peter M. Luthy, Guido L. Weiss, and Steven S. Xiao
August 2016

CHAPTER 1

Limits, Continuity, and Compactness

1.1 NUMBER SYSTEMS AND THE PRINCIPLE OF MATHEMATICAL INDUCTION

We state above that you are familiar with the natural numbers $\mathbb{N} = \{1, 2, 3, \cdots\}$. These are the numbers we use for counting. A very important tool associated with $\mathbb{N}$ is the *Principle of Mathematical Induction*, which we now present and explain to you. Suppose we want to establish that a fact holds for every natural number $n \in \mathbb{N}$. Let us denote this fact by $P(n)$. For example, let $P(n)$ be the statement that $5^n - 1$ is a multiple of 4 for each $n \in \mathbb{N}$. One can easily verify that $P(1), P(2), P(3)$, and $P(4)$ are all true by making some obvious calculations. Also observe that, if $5^n - 1$ is a multiple of 4, then $4 \cdot 5^n + (5^n - 1) = 5^{n+1} - 1$ is also a multiple of 4. This is equivalent to the statement that, if $P(n)$ is a multiple of 4, $P(n + 1)$ must also be a multiple of 4. Thus, since we have verified that $P(1)$ is true, it must be the case that $P(2)$ is true. Since $P(2)$ is true, it must be the case that $P(3)$ is true, and so on. This pattern continues endlessly, and so $P(n)$ must be true for any $n \in \mathbb{N}$.

The Principle of Mathematical Induction is the following general fact: if $P(n)$ is a family of statements, one for each n, and we can establish that

(a) $P(1)$ is true.

(b) If $P(n)$ is true, then $P(n + 1)$ is true.

then $P(n)$ must be true for all $n \in \mathbb{N}$.

We shall discuss other examples of this principle. Let us begin with an often told story about the eminent 18th-century mathematician Gauss when he was a nine-year-old boy. One day Gauss misbehaved in a class he was attending. The teacher sent him to a corner of the classroom and told him to sum the first one hundred natural numbers, $1 + 2 + \cdots + 100 = \sum_{k=1}^{100} k$. By the time Gauss reached the corner, he blurted out, "the sum is 5,050". He was asked how he obtained the answer so quickly. He said he lined up the sum twice, as follows:

$$
\begin{array}{ccccccccccc}
1 & + & 2 & + & 3 & + & \cdots & + & 98 & + & 99 & + & 100 \\
100 & + & 99 & + & 98 & + & \cdots & + & 3 & + & 2 & + & 1
\end{array}
$$

Each column sums to 101, and there are 100 columns, so the sum of all the numbers in the two lines is 101×100. This is exactly twice as large as $1 + 2 + 3 + \cdots + 100$, so the answer is

$$\frac{101 \times 100}{2} = 5{,}050.$$

Of course, this sets up the following more general problem: if $\sum_{k=1}^{n} k$, then we expect the sum to be $\dfrac{n(n + 1)}{2}$ based on Gauss's solution. If, for each $n \in \mathbb{N}$, we let $P(n)$ be the statement that $\sum_{k=1}^{n} k = \dfrac{n(n + 1)}{2}$, then we can easily prove $P(n)$ holds for all $n \in \mathbb{N}$ by induction:

(a) $P(1)$ is simply the assertion that $\sum_{k=1}^{1} k = 1 = \dfrac{1}{1 + 1} \cdot 2$, which is of course true.

(b) If we assume $P(n)$ is true, that is, $\sum_{k=1}^{n} k = \dfrac{n(n + 1)}{2}$. Then

$$\sum_{k=1}^{n+1} k = \left(\sum_{k=1}^{n} k \right) + n + 1 = \frac{n(n + 1)}{2} + n + 1 = \frac{n^2 + 3n + 2}{2} = \frac{(n + 1)(n + 2)}{2}.$$

That the first and last terms of this string of equalities are equal is precisely the statement that $P(n + 1)$ is true.

This last example can be extended into a much larger family of identities. In the last example, we started with a sum, $\sum_{k=1}^{n} k$, which we now denote by $p_1(n)$. We also found a polynomial, $\dfrac{n(n + 1)}{2}$, which we denote by $s_1(n)$. We found that the claim "$p_1(n) = s_1(n)$ for all $n \in \mathbb{N}$" satisfied the conditions of the mathematical induction.

Let us now define the following: For each $\ell \in \{0, 1, 2, \cdots\}$, let

$$p_\ell(n) = \sum_{k=1}^{n} k^\ell.$$

Thus $p_\ell(n)$ denotes the sum of the first n "ℓ-th powers". The analogous question asked of Gauss is to sum the first 100 ℓ-th powers. Gauss's previous method will not work. Our goal is to find a polynomial $s_\ell(n)$ which equals $p_\ell(n)$. Before continuing on, we encourage the reader to complete the following exercise.

(HW) 1.1 In this problem, we ask you to mimic the induction methodology we used to prove that $p_1(n) = \dfrac{n(n+1)}{2} = s_1(n)$ to establish the following.

(a) What is a natural choice for $s_0(n)$? Use induction to prove the statement "$p_0(n) = s_0(n)$ for every $n \in \mathbb{N}$".

(b) Let $s_2(n) = \dfrac{n(n+1)(2n+1)}{6}$. Use induction to prove that $p_2(n) = s_2(n)$ for every $n \in \mathbb{N}$.

With part (b) of the previous exercise in mind, the reader may well wonder how one could come up with the formula for $s_\ell(n)$—in applying induction, one certainly needed to know this formula! Here is a first approach—we will give a more robust approach a little bit later.

Suppose we wish to find a simple expression for $p_3(n)$—that is we want to find a simple formula for $s_3(n)$. We first observe that $s_0(n)$ was a degree one polynomial, $s_1(n)$ was a degree two polynomial, and $s_2(n)$ was a degree three polynomial. So we may well guess that $s_3(n)$ is a degree four polynomial so that $s_3(n) = an^4 + bn^3 + cn^2 + dn + e$. Thus we claim we can find coefficients a, b, c, d, and e so that

$$\sum_{k=1}^{n} k^3 = an^4 + bn^3 + cn^2 + dn + e.$$

We can find out what a, b, c, d, and e are as follows: assume the equality holds and plug in $n = 1, 2, 3, 4, 5$. This produces the system of equations,

$$
\begin{array}{rcllllll}
1 & = & a & + & b & + & c & + & d & + & e \\
9 & = & 16\,a & + & 8\,b & + & 4\,c & + & 2\,d & + & e \\
36 & = & 81\,a & + & 27\,b & + & 9\,c & + & 3\,d & + & e \\
100 & = & 128\,a & + & 64\,b & + & 16\,c & + & 4\,d & + & e \\
225 & = & 625\,a & + & 125\,b & + & 25\,c & + & 5\,d & + & e.
\end{array}
$$

We assume you are capable of solving such a system, though it might be a bit laborious. Doing so yields $a = \dfrac{1}{4}, b = \dfrac{1}{2}, c = \dfrac{1}{4}$, and $d = e = 0$. So, our guess is that

$$s_3(n) = \frac{1}{4}n^4 + \frac{1}{2}n^3 + \frac{1}{4}n^2 = \frac{n^2(n+1)^2}{4}.$$

With this answer in hand, it is easy to check that the assertion "$p_3(n) = s_3(n)$ for all $n \in \mathbb{N}$" satisfies the condition for induction:

(a) $p_3(1) = 1^2 = 1$ and $s_3(1) = \dfrac{1^2 \cdot 2^2}{4} = 1$.

(b) If for some $n \in \mathbb{N}$, we know that $p_3(n) = s_3(n)$, that is $\sum_{k=1}^{n} k^3 = \dfrac{n^2(n+1)^2}{4}$, then

$$p_3(n+1) \;=\; \sum_{k=1}^{n+1} k^3 = \left(\sum_{k=1}^{n} k^3\right) + (n+1)^3 = \dfrac{n^2(n+1)^2}{4} + (n+1)^3$$

$$= \dfrac{(n+1)^2(n+2)^2}{4},$$

where the last step involves some simple algebra. The right side of this equality is, of course, equal to $s_3(n+1)$, so we conclude that if $p_3(n) = s_3(n)$, then $p_3(n+1) = s_3(n+1)$.

One could apply this same approach to finding a formula for $s_4(n)$: assume $s_4(n)$ is a degree five polynomial so that $p_4(n) = a n^5 + b n^4 + c n^3 + d n^2 + e n + f$, plug in $n = 1, 2, 3, 4, 5, 6$ and solve the system of equations this generates to find the coefficients, and finally apply induction to verify that the formula is, indeed, correct. However, the system of equations approach is an ad hoc method—"ad hoc" is a Latin phrase meaning "for a particular purpose". The system of equations approach is ad hoc in the sense that being able to determine and solve the system of equations for $\ell = 3$ does not really help one determine and solve the system for $\ell = 4$. We now give a much more robust method for determining formulas for s_ℓ.

Suppose we wished to find s_2. One can easily check that for every $j \in \mathbb{N}$, $(j+1)^3 - j^3 = 3 j^2 + 3 j + 1$. Use this identity to observe the following:

$$
\begin{aligned}
2^3 - 1^3 &= 3 \cdot 1^2 + 3 \cdot 1 + 1 \\
3^3 - 2^3 &= 3 \cdot 2^2 + 3 \cdot 2 + 1 \\
4^3 - 3^3 &= 3 \cdot 3^2 + 3 \cdot 3 + 1 \\
\cdots &= \cdots \\
(n+1)^3 - n^3 &= 3 \cdot n^2 + 3 \cdot n + 1.
\end{aligned}
$$

Adding all these equations together yields

$$(n+1)^3 - 1 = 3 \, p_2(n) + 3 \, p_1(n) + n.$$

We can solve for $p_2(n)$ to produce

$$p_2(n) = \frac{1}{3} [(n+1)^3 - 1 - 3 \, p_1(n) - n].$$

We have the formula, $p_1(n) = \dfrac{n(n+1)}{2}$, so

$$p_2(n) = \frac{1}{3}\left[(n+1)^3 - 1 - 3\,\frac{n(n+1)}{2} - n\right] = \frac{n(n+1)(2n+1)}{6}.$$

This argument can be extended to all $\ell \in \mathbb{N}$ as follows. There is a general theorem called the Binomial Theorem which enables one to compute $(x + 1)^\ell$ for any ℓ. For example,

$$\begin{aligned}
(x + 1)^4 &= x^4 + 4x^3 + 6x^2 + 4x + 1 \\
(x + 1)^5 &= x^5 + 5x^4 + 10x^3 + 10x^2 + 5x + 1 \\
(x + 1)^6 &= x^6 + 6x^5 + 15x^4 + 20x^3 + 15x^2 + 6x + 1 \\
(x + 1)^7 &= x^7 + 7x^6 + 21x^5 + 35x^4 + 35x^3 + 21x^2 + 7x + 1,
\end{aligned}$$

and so one can write

$$\begin{aligned}
(x + 1)^4 - x^4 &= 4x^3 + 6x^2 + 4x + 1 \\
(x + 1)^5 - x^5 &= 5x^4 + 10x^3 + 10x^2 + 5x + 1 \\
(x + 1)^6 - x^6 &= 6x^5 + 15x^4 + 20x^3 + 15x^2 + 6x + 1 \\
(x + 1)^7 - x^7 &= 7x^6 + 21x^5 + 35x^4 + 35x^3 + 21x^2 + 7x + 1.
\end{aligned}$$

To find $p_3(n)$, one need only now use the first of these identities and mimic the previous method:

$$\begin{aligned}
2^4 - 1^4 &= 4 \cdot 1^3 + 6 \cdot 1^2 + 4 \cdot 1 + 1 \\
3^4 - 2^4 &= 4 \cdot 2^3 + 6 \cdot 2^2 + 4 \cdot 2 + 1 \\
4^4 - 3^4 &= 4 \cdot 3^3 + 6 \cdot 3^2 + 4 \cdot 3 + 1 \\
\cdots &= \cdots \\
(n + 1)^4 - n^4 &= 4 \cdot n^3 + 6 \cdot n^2 + 4 \cdot n + 1.
\end{aligned}$$

Add the columns to produce

$$(n + 1)^4 - 1 = 4\, p_3(n) + 6\, p_2(n) + 4\, p_1(n) + n,$$

and solve for $p_3(n)$:

$$p_3(n) = \frac{1}{4}[(n + 1)^4 - 1 - 6\, p_2(n) - 4\, p_1(n) - n] = \frac{1}{4} n^2 (n + 1)^2.$$

This approach of finding $s_\ell(n)$ is a *recursive* method: finding the formula for $s_{30}(n)$ is reduced to knowing formulas for $s_1(n), s_2(n), \cdots, s_{29}(n)$. But the formula for $s_{29}(n)$ can be found for knowing $s_1(n), s_2(n), \cdots, s_{28}(n)$, and the formula for $s_{28}(n)$ can be found for knowing $s_1(n), s_2(n), \cdots, s_{27}(n)$, and so on. Recursion can be an extremely powerful problem-solving method.

(HW) 1.2   Obtain the formula for $s_4(n)$, where $\displaystyle\sum_{k=1}^{n} k^4 = s_4(n)$ and $s_4(n)$ is a polynomial.

(HW) 1.3   Explain how to obtain formula for $\displaystyle\sum_{k=1}^{n} k^\ell$ for $\ell > 4$.

(HW) 1.4   Show that, for every $n \in \mathbb{N}$,

$$\frac{1^2 + 2^2 + \cdots + n^2}{1 + 2 + \cdots + n} = \frac{2n + 1}{2}.$$

(Hint: $\dfrac{1^2}{1} = \dfrac{3}{3}, \dfrac{1^2 + 2^2}{1 + 2} = \dfrac{5}{3}, \dfrac{1^2 + 2^2 + 3^2}{1 + 2 + 3} = \dfrac{7}{3}, \cdots$). Use this result for obtaining another proof of sum of squares?

(HW) 1.5

(a) Use the Principle of Mathematical Induction to prove that

$$\sum_{k=1}^{n} (2k - 1) = n^2$$

   for all positive integers n.

(b) Provide a different proof of the identity in part (a) by splitting the sum up and using the fact
   that $\dfrac{n(n + 1)}{2} = 1 + 2 + \cdots + n$.

(HW) 1.6   This exercise references the previous exercise. In this problem, we consider the following pattern in the odd numbers:

- $1 = 1^3$.

- Add the next two odds: $3 + 5 = 8 = 2^3$.

- Add the next three odds: $7 + 9 + 11 = 27 = 3^3$.

- Add the next four odds: $13 + 15 + 17 + 19 = 64 = 4^3$.

(a) First, write down a conjecture of the form

$$n^3 = \sum_{k=f(n)}^{f(n+1)-1} (2k - 1)$$

   for all $n \geq 1$ and make the function $f(n)$ explicit. (Hint: count how many odd numbers you skip in the above pattern.)

(b) Based on your conjecture above, write down a second conjecture of the form

$$\sum_{k=1}^{n} k^3 = \sum_{k=1}^{f(n+1)-1} (2k-1).$$

(c) Prove the second conjecture. (Hint: use the formula from the previous problem.)

(d) Deduce from this that the first conjecture is true.

(e) Based on these computations, what is the relationship between

$$1^3 + 2^3 + \cdots + n^3$$

and

$$1 + 2 + \cdots + n\,?$$

1.2   A QUICK INTRODUCTION TO CARDINAL NUMBERS

For a finite set A, we define the *cardinality* of A to be the number of elements of A. For example, if $A = \{1, 3, \pi\}$, then the cardinality of A is 3. Cardinality gives us a notion of "how big" a finite set is. Many sets are not of finite cardinality: $\mathbb{N}$, $\mathbb{Z}$, $\mathbb{Q}$, and $\mathbb{R}$ each contain infinitely many elements.

There is another way to consider "how big" finite sets are relative to one another. Let A and B be sets. Recall that a function $f : A \to B$ is injective if and only if $f(a) = f(a')$ implies $a = a'$. Recall also that $f : A \to B$ is surjective if and only if, for every $b \in B$, there exists an $a \in A$ such that $f(a) = b$. In other words, B is the range of f. A function is bijective if it is both injective and surjective. If A and B are both finite sets, then they have the same number of elements if and only if there exists a function $f : A \to B$ which is bijective. We use this notion to extend the notion of cardinality to sets with infinitely many elements.

Definition 1.7   *Given two sets A and B (which can contain either finitely or infinitely many elements) we say that A and B have the same cardinality if and only if there exists a bijection $f : A \to B$.*

It may be helpful to call a function $f : A \to B$ that is bijective to be *one to one* and *onto*.

One is led to the following question: do all sets with infinitely many elements have the same cardinality? This is precisely the question the mathematical community became interested in during the end of the 19th century, following the seminal and revolutionary discoveries of the German mathematician Georg Cantor.

Let us begin by pointing out that, for infinite sets, the notion of cardinality is quite different from that of finite sets. For example, if A and B are both finite, $A \subset B$, and B contains an element which is not an element of A, then there cannot be a bijection between A and B, and so A and B have different cardinality. However, consider the sets $\mathbb{N}$ and $E := \{2, 4, 6, 8, \cdots\}$—of course,

E is the set of positive, even integers. Let $f : \mathbb{N} \to E$ be defined by $f(n) = 2n$. This function is certainly surjective, as every positive, even integer is, by definition, two times a positive integer. Moreover, f is injective: if $f(n) = f(n')$, then $2n = 2n'$, and, by elementary arithmetic, $n = n'$. Thus f is a bijection and $\mathbb{N}$ and E have the same cardinality. More generally, if A and B are any sets with infinitely many elements such that $A \subset B$, it could be the case that A and B have the same cardinality.

(HW) 1.8   Give an example of a bijection between $\mathbb{N}$ and $\mathbb{Z}$.

Could it be the case that $\mathbb{N}$ and $\mathbb{Q}$ have the same cardinality? It seems "obvious" that $\mathbb{Q}$ is "bigger" than $\mathbb{N}$. Yet one can produce a bijection between $\mathbb{N}$ and $\mathbb{Q}$!

Theorem 1.9 $\mathbb{N}$ *and* $\mathbb{Q}$ *have the same cardinality.*

Proof: It is not such a hard exercise to show that the set of positive rational numbers, $\mathbb{Q}^+$, has the same cardinality as $\mathbb{Q}$. Because the composition of two bijections is still a bijection, it suffices to prove that $\mathbb{N}$ and $\mathbb{Q}^+$ have the same cardinality. We leave it as an exercise following this proof. Arrange the positive rational numbers as an infinite matrix whose first row consists of the numbers of $\mathbb{N}$, whose second row consists of those positive fractions whose denominators are 2, whose third row consists of those positive fractions whose denominators are 3, and so on. See Fig. 1.1.

To create the bijection, we create a path by moving back and forth along the diagonals, as indicated by the arrows in the diagram; skip a rational in the matrix if it is equivalent to a fraction already crossed by our path (the circled points are the skipped ones). The bijection is then $g(1) = 1$, $g(2) = 2$, $g(3) = \frac{1}{2}$, $g(4) = \frac{1}{3}$, $g(5) = 3$, $g(6) = 4$, $g(7) = \frac{3}{2}$, $g(8) = \frac{2}{3}$, $g(9) = \frac{1}{4}$, $g(10) = \frac{1}{5}$, $g(11) = 5, \cdots$.

By construction, it is clear that this function g is injective since equivalent fractions are never repeated. Given any two positive integers, a and b, with $b \neq 0$, the path given in the diagram will eventually cross the fraction $\frac{a}{b}$, and so the function g is also surjective. Thus g is bijective, whence $\mathbb{N}$ and $\mathbb{Q}^+$ have the same cardinality.   ∎

(HW) 1.10   Complete the proof of the above theorem by proving that $\mathbb{Q}$ and $\mathbb{Q}^+$, the set of positive rational numbers, have the same cardinality.

Definition 1.11   *A set A such that a bijection $f : \mathbb{N} \to A$ exists is said to have cardinality* $\aleph_0$.[1]

Any set which is of finite cardinality or whose cardinality is $\aleph_0$ is said to be *countable*. A natural question is whether or not uncountable sets exist. Cantor answered this question in the affirmative. He showed that uncountable sets exist and that $\mathbb{R}$, in particular, is uncountable.

---

[1] $\aleph$ is the first letter, aleph, of the Hebrew alphabet. $\aleph_0$ is read as "aleph naught".

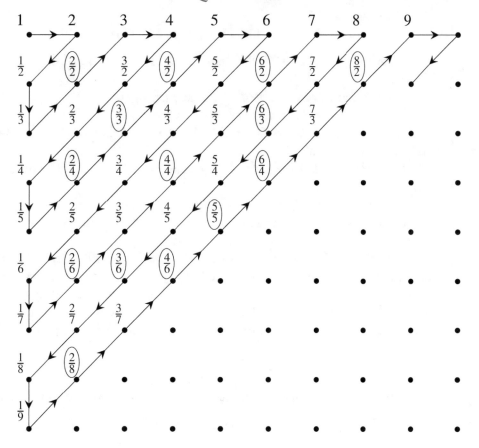

Figure 1.1: A path in the non-negative rationals.

Let us first consider the notion of power set.

Definition 1.12   *Let A be a set. The symbol 2^A denotes the power set of A, which is the set of all subsets of A. That is, $2^A := \{S : S \subset A\}$.*

For example, if $A = \{1, 2, 3\}$, then $2^A = \{\varnothing, \{1\}, \{2\}, \{3\}, \{1, 2\}, \{1, 3\}, \{2, 3\}, \{1, 2, 3\}\}$. For any finite set A, if A consists of n elements, then 2^A consists of 2^n elements (which explains the notation 2^A)—in particular, 2^A is always a set of strictly greater cardinality. The following result asserts that this property extends to infinite sets.

Theorem 1.13   *Suppose that A is a set. Then there is no bijection between A and 2^A.*

Proof: For any function $f : A \to 2^A$. We claim that f cannot be surjective.

Given an $a \in A$, let

$$D := \{a \in A : a \notin f(a)\}.$$

Clearly $D \subset A$ so $D \in 2^A$. We claim that D is not in the range of f: that is, for any $a \in A$, $f(a) \neq D$.

Suppose there exists an $x \in A$, such that $f(x) = D$.

There are two possibilities: either $x \in f(x)$ or $x \notin f(x)$.

If $x \in f(x)$, $x \in D \ (= f(x))$. But this, by the definition of D, means $x \notin f(x) = D$ which is a contradiction!

If $x \notin f(x)$, $x \notin D \ (= f(x))$. By the definition of D, $x \in f(x)$ which contradicts to $x \notin f(x)$.

Hence, f is not surjective and, in particular, not a bijection.

Since f was an arbitrary function from A to 2^A, we conclude that there is no bijection between A and 2^A. ∎

Finally, let us show that the reals are not countable. We will consider $\mathbb{N}$ and $[0, 1]$. Elements of $[0, 1]$ shall be represented by their decimal expansions. We will discuss the notion of limits shortly, but, as we stated above, we assume you know enough about this notion that you can understand the decimals for $[0, 1]$: a real number x has the property that $x \in [0, 1]$ if and only if its decimal expansion is $x = 0.a_1 a_2 a_3 \cdots a_n \cdots$, where each $a_j \in \{0, 1, \cdots, 8, 9\}$. This means that

$$x = \frac{a_1}{10} + \frac{a_2}{10^2} + \cdots + \frac{a_n}{10^n} + \cdots.$$

This series converges and it uniquely represents x except for equalities of the following form:

$$0.a_1 a_2 \cdots a_{n-1} a_n 99999 \cdots = 0.a_1 a_2 \cdots a_{n-1}(a_n + 1)0000 \cdots,$$

where $a_n < 9$, and $1.00000 \cdots = 0.999999 \cdots$. The only numbers in $[0, 1]$ which can be expressed by decimals in more than one way can be expressed by a decimal which ends with an infinite string of consecutive 9's. Such a number equals one that ends in infinite many 0's (as indicated a few lines up). To avoid confusion, if a number $x \in [0, 1]$ has more than one decimal expansion, we will only use the (unique) decimal expansion which ends in a string of 9's.

Theorem 1.14 $[0, 1]$ *is an uncountable infinite set.*

Proof: Suppose that $f : \mathbb{N} \to [0, 1]$ were a bijection. For each $k \in \mathbb{N}$, the real number $f(k)$ has a decimal expansion of the form $0.a_1^k a_2^k \cdots$ (here a_i^k does not denote a_i raised to the kth power but rather signifies that it corresponds to a decimal digit for $f(k)$).

Consider the nth digit of $f(n)$, i.e., a_n^n. Define a new digit b_n so that if $a_n^n \neq 4$, then $b_n = 4$, and, if $a_n^n = 4$, then $b_n = 5$. This means that $b_n \neq a_n^n$. Consider the decimal number $b := 0.b_1 b_2 \cdots$. Certainly, b corresponds to a number in $[0, 1]$. We claim that b is not in the range of f. Suppose there were some integer m so that $f(m) = b$. This means that $a_i^m = b_i^m$ for every $i \in \{1, 2, \cdots\}$ since, as per the discussion just prior to the statement of this theorem, our decimal expansions uniquely determine elements of $[0, 1]$. However, we defined b so that $b_m \neq a_m^m$, so b and $f(m)$ must actually differ at the mth decimal digit, which is a contradiction. This means that b is not in the range of f, and so f cannot be a bijection.   ■

This proof is generally referred to as a "Cantor diagonalization argument", paying homage to its creator, Cantor, as well as to the fact that the diagonal, a_n^n plays a prominent role.[2]

Theorem 1.15   *The cardinality of $\mathbb{R}$ is different from $\aleph_0$; in particular, $\mathbb{R}$ is uncountable.*

(HW) 1.16   Give a bijection between $[0, 1]$ and $\mathbb{R}$, then use this bijection and Theorem 1.14 to prove Theorem 1.15.

One can prove that, actually, $2^{\mathbb{N}}$ and $\mathbb{R}$ have the same cardinality, though we will not prove it here. A natural question is whether there are any sets which are "bigger" than $\mathbb{N}$ but "smaller" than $\mathbb{R}$, in terms of cardinality. We say that a set A *has larger cardinality than* B if there is an injective function from B to A. If there is an injection from A to B and an injection from B to A, then there is a bijection between A and B (this fact is known as the Schroeder-Bernstein Theorem), and thus A and B have the same cardinality. So, A is said to *have strictly smaller cardinality than* B if there is an injection from A to B but no injection from B to A. Cantor conjectured that there is no set A with the property that $\mathbb{N}$ has strictly smaller cardinality than A and the property that A has strictly smaller cardinality than $\mathbb{R}$. However, Cantor was unable to prove this. His conjecture became known as the Continuum Hypothesis and resisted proof for nearly a century. In 1963, Paul Cohen proved that, actually, the Continuum Hypothesis is independent from the axioms of set theory. In short, what Cohen proved is that there are actually two equivalently valid mathematical universes: one where the Continuum Hypothesis is true and one where it is false. So, it is not surprising, in retrospect, that no one could prove the Continuum Hypothesis.

1.3   LIMITS

Let us present some more properties of the real numbers. The real numbers are made up of the positive reals $\mathbb{R}^+$, the negative reals $\mathbb{R}^-$ and $\{0\}$. These sets allow us to define the order relation less than, $<$, that applies to each pair $x, y \in \mathbb{R}$: if $x \neq y$, we then write $x < y$ if and only if $y - x \in \mathbb{R}^+$ which is often written as $y - x > 0$. In a similar way, we write $y < x$ if and only if $y - x \in \mathbb{R}^-$. Observe that this relation is *transitive*: $x < y$ and $y < z$ implies $x < z$.

The real numbers form a *field* with respect to addition and multiplication, which we write as $+$ and $\cdot$. (We assume you are fairly comfortable with addition and multiplication of decimals). Given $a, b \in \mathbb{R}$, their sum and product are also real numbers, i.e., $a + b \in \mathbb{R}$ and $a \cdot b \in \mathbb{R}$. Both operators are *associative*: $(a + b) + c = a + (b + c)$ and $(a \cdot b) \cdot c = a \cdot (b \cdot c)$. They are also *commutative*: $a + b = b + a$ and $a \cdot b = b \cdot a$. Addition has an *identity element*, 0, and multiplication has an *identity element*, 1: $0 + x = x$ and $1 \cdot x = x$. The *distribution law* also holds:

---

[2]The proof we gave previously that there is no bijection between A and 2^A is also an example of a diagonalization argument.

$a \cdot (b + c) = a \cdot b + a \cdot c$. The equation $a + x = b$ has a solution for every $a, b \in \mathbb{R}$, and the equation $a \cdot x = b$ has a solution for all $a, b \in \mathbb{R}$ provided $a \neq 0$. When $b = 0$ we write the solution to $a + x = 0$ as $-a$ and the solution to $a \cdot x = 1$ (when $a \neq 0$) as a^{-1} or $\dfrac{1}{a}$.

(HW) 1.17   Show that the additive and multiplicative identity elements are unique. Show that the additive and multiplicative inverse elements $-a$ and a^{-1} (when $a \neq 0$) are uniquely determined by a.

The rationals, $\mathbb{Q}$, also form a field, a subfield of $\mathbb{R}$. Of course, neither $\mathbb{N}$ nor $\mathbb{Z}$ are fields. For example, multiplicative inverse is non-existant for most elements. The order relation $<$ makes perfect sense in these subsets of $\mathbb{R}$. The same is true for $\leq$ and $\geq$. Let us say a few more words about these relations.

A subset S of $\mathbb{R}$ is *bounded above* iff there exists $N \in \mathbb{R}$, such that $x \leq N$ for all $x \in S$; it is *bounded below* iff $-S = \{-y : y \in S\}$ is bounded above. S is *bounded* iff it is both bounded above and below. A subset of the positive integers is always bounded below (since $1 \leq n$ for all $n \in \mathbb{N}$). s is a *least element* of S iff $s \leq x$ for all $x \in S$ and $s \geq t$ whenever $t < x$ for all $x \in S$. Not every subset S of $\mathbb{R}$ contains a least element, even when it is bounded below (think of some examples). However, every subset of $\mathbb{N}$ contains a least element as long as it is nonempty—this is called the *Well-Ordering Principle*.

Suppose $S \subset \mathbb{R}$ and S is bounded above. Thus, there exists $b \in \mathbb{R}$ such that $s \leq b$ for all $s \in S$. There may be a smaller upper bound of S. Is there a smallest one? That is, does there exist $b_0 \in \mathbb{R}$ satisfying $s \leq b_0$ for all $s \in S$ and, whenever b is an upper bound of S, we have $b_0 \leq b$? For the real numbers, the answer is yes: there is such a $b_0 \in \mathbb{R}$. It is clearly unique and is called *the least upper bound* of S; it is denoted by LUB (of S). It is also called the *supremum* of S and is denoted by $sup\, S$. If $sup\, S \in S$ it is called the *maximum* of S and is denoted by $max\, S$. We emphasize this important property of $\mathbb{R}$:

(*)   *Each $S \subset \mathbb{R}$ which is bounded above has a LUB that belongs to $\mathbb{R}$.*

The term *complete* is used to denote (*). In fact, (*) characterizes $\mathbb{R}$.

(HW) 1.18   Show that $\sqrt{2}$ cannot be rational.

This fact is, perhaps, the most obvious one that shows that $\mathbb{Q}$ is not complete. There are many examples, like e and π, that show that $\mathbb{Q}$ is not complete. The material we discuss (very soon) about sequences is needed for establishing these facts about e and π.

Let us announce a few more properties of the order relation $\geq$:

• If $b \geq c$, then $a + b \geq a + c$ for all $a \in \mathbb{R}$.

• $1 > 0$.

• if $b \geq c$ and $a \geq 0$, then $a \cdot b \geq a \cdot c$.

• If $a > 0$ and $b > 0$, there exists $n \in \mathbb{N}$ such that $n \cdot a > b$.

This is called the *Archimedian* property of $\geq$. A field with an order relation satisfying these properties is called an Archimedian ordered field. $\mathbb{Q}$ and $\mathbb{R}$ are Archimedian ordered fields. $\mathbb{Q}$ is not complete.

Let us introduce the notion of a *limit*. We shall see how this notion is connected to the above properties. The subject called "the Calculus" is based on the notion of limit. Most important, as we shall see, calculus is really defined by two limits: the derivative and the integral.

We begin with the notion of a *sequence* $\{s_n\}$ which is a function f on $\mathbb{N}$ into a set S: $f : \mathbb{N} \to S$ and the notation we use is $f(n) = s(n)$ or $s_n, n \in \mathbb{N}$. Let us talk about sequences in the real line when $S = \mathbb{R}$: We say $\{s_n\}, n \in \mathbb{N}$, *converges* to a real number s iff for each $\epsilon > 0$, there exists $N = N(\epsilon)$, such that $|s_n - s| < \epsilon$ if $n \geq N$ (the *absolute value* of $r \in \mathbb{R}$, denoted by $|r|$, is r if $r \geq 0$ and is $-r$ otherwise). The number s is called the *limit* of $\{s_n\}$. When this is the case we write

$$\lim_{n \in \mathbb{N}} s_n = \lim_{n \to \infty} s_n = s.$$

(HW) 1.19   Show that if $s = \lim s_n$ exists, then s is unique.

If $s = 0$, $\{s_n\}$ is called a *null sequence*.

(HW) 1.20

(a) Show that $\{s_n\} = \{\dfrac{1}{n}\}, n \in \mathbb{N}$, is a null sequence.

(b) Suppose $0 < a < 1$, show that $\{a^n\}, n \in \mathbb{N}$, is a null sequence.

Definition 1.21   *A sequence $\{s_n\}$ is a Cauchy sequence iff for each $\epsilon > 0$, there exists $N = N(\epsilon) \in \mathbb{N}$ such that $|s_n - s_m| < \epsilon$ if $n, m \geq N(\epsilon)$.*

This last property leads us to a natural question: Does a Cauchy sequence in $\mathbb{R}$ have a limit in $\mathbb{R}$? The answer is "yes" and we have the following result:

Theorem 1.22   *A sequence $\{s_n\}$ has a limit in $\mathbb{R}$ iff it is Cauchy.*

The proof will be an easy consequence of several results we will drive from the LUB property that will give us a better understanding of the notion of completeness and its connection with order in $\mathbb{R}$.

We call a sequence $\{s_k\}, k \in \mathbb{N}$, *monotone* if and only if it satisfies $s_k \leq s_{k+1}$ for $k \in \mathbb{N}$, or $s_n \geq s_{k+1}$ for $k \in \mathbb{N}$. In the first case $\{s_k\}$ is called *monotone increasing* (or, simply *increasing*) and, in the second case, *monotone decreasing* (or, simply, *decreasing*).[3]

Let us suppose $\{s_k\}$ is a bounded sequence so that it has a GLB a_0 and a LUB b_0. We then have, for all $k \in \mathbb{N}$, $a_0 \leq s_k \leq b_0$, and let us assume $b_0 - a_0 > 0$ (so there are infinitely many elements s_k of the sequence). Let $I_0 = [a_0, b_0]$. Either $[a_0, \dfrac{a_0 + b_0}{2}]$ or $[\dfrac{a_0 + b_0}{2}, b_0]$ have infinitely many elements of $\{s_k\}$. Select one of these intervals that contain infinitely many elements of $\{s_k\}$ and denote it by $[a_1, b_1]$. Clearly one of the intervals $[a_1, \dfrac{a_1 + b_1}{2}]$ and $[\dfrac{a_1 + b_1}{2}, b_1]$ contain infinitely many of the s_k's and denote it $[a_2, b_2]$. We repeat this procedure for each $n \in \mathbb{N}$, we obtain a sequence of closed intervals $I_n = [a_n, b_n]$, $n \in \mathbb{N} \cup \{0\}$, satisfying

(i) $I_{n+1} \subset I_n$, $n \in \mathbb{N} \cup \{0\}$;

(ii) the lengths $|I_n| = b_n - a_n$ satisfy $|I_{n+1}| = 2^{-(n+1)}|I_0|$;

(iii) $[a_{n+1}, b_{n+1}]$ is either $[a_n, \dfrac{a_n + b_n}{2}]$ or $[\dfrac{a_n + b_n}{2}, b_n]$.

The sequence $\{a_n\}$ is *monotone increasing* ($a_0 \leq a_1 \leq a_2 \leq \cdots \leq a_n \leq \cdots$) and the sequence $\{b_n\}$ is *monotone decreasing*, each is bounded below by a_0 and above by b_0. It follows that each of these sequences is convergent.

The following two results are easily proved by the observations we just made.

Lemma 1.23 *Any bounded monotone sequence has a limit.*

Theorem 1.24 *If $\{s_n\}$ is a bounded sequence, then there exists a convergent subsequence $\{s_{k_n}\}$, $n \in \mathbb{N}$.*

We are now ready to prove Theorem 1.22.
First of all it is easy to show that $\{s_n\}$ is Cauchy implies $\{s_n\}$ is bounded.

(HW) 1.25 Prove this last fact.

It is immediate that if a sequence has a limit, then it is Cauchy.

It is clear from the above observations that a Cauchy sequence is bounded. From this and Theorem 1.24 we obtain Theorem 1.22.

(HW) 1.26 Show that if $\lim\limits_{n} s_n = s$ and $\lim\limits_{n} t_n = t$, then $\lim\limits_{n}(s_n + t_n) = s + t$ and $\lim\limits_{n} s_n \cdot t_n = s \cdot t$.

---

[3]If the symbol $<$, instead of $\leq$, is used we say $\{s_k\}$ is *strictly increasing*, if $>$, instead of $\geq$, is used, $\{s_k\}$ is *strictly decreasing*.

We have been considering limits of sequences of real numbers. A sequence $\{s_n\}$, $n \in \mathbb{N}$, is a function $f : \mathbb{N} \to \mathbb{R}$, $s_n = f(n)$ for $n \in \mathbb{N}$. More generally, we can consider a function whose domain is a subset S of $\mathbb{R}$ that maps S into $\mathbb{R}$: $f : S \to \mathbb{R}$. Suppose that x_0 is a limit point of S (this means that each open interval about x_0, $(s_0 - \frac{\epsilon}{2}, x_0 + \frac{\epsilon}{2})$, of length $\epsilon > 0$ contains a point of S). We introduce the notion of the *limit of f at x_0*: this is a number ℓ such that for each $\epsilon > 0$ there exists $\delta > 0$ for which $|f(x) - \ell| < \epsilon$ if $x \in S$ satisfies $|x - x_0| < \delta$. This notion can be reduced to a statement involving sequential limits.

Theorem 1.27   *Let $f : S \to \mathbb{R}, S \subset \mathbb{R}$, and suppose x_0 is a limit point of S. Then ℓ is a limit point at x_0 ($\lim\limits_{x \to x_0, x \in S} f(x) = \ell$) if and only if whenever $\{x_n\}$ is a sequence of points in S such that $\lim\limits_{n} x_n = x_0$ then $\lim\limits_{n} f(x_n) = \ell$.*

(HW) 1.28   Prove this theorem.

1.4   VECTOR SPACE, METRIC SPACE, NORMS, AND INEQUALITIES

Most of the notions and results we have stated have an extension to more general settings: *vector spaces*. $\mathbb{R}$ is a special (one dimensional) *vector space V*.

Definition 1.29   *A vector space V is a non-empty set V of objects, called vectors, on which are defined two operations, called addition and multiplication by scalars (real numbers), subject to the ten axioms (or rules) listed below.*

1. *The sum of u and v, denoted by $u + v$, is in V.*

2. *$u + v = v + u$.*

3. *$(u + v) + w = u + (v + w)$.*

4. *There is a zero vector 0 in V such that $u + 0 = u$.*

5. *For each u in V, there is a vector $-u$ in V such that $u + (-u) = 0$.*

6. *The scalar multiple of u by α, denoted by αu, is in V.*

7. *$\alpha(u + v) = \alpha u + \alpha v$.*

8. *$(\alpha + \beta)u = \alpha u + \beta u$.*

9. *$\alpha(\beta u) = (\alpha\beta)u$.*

10. $1u = u$.

for all elements $u; v, w \in V$ *and* $\alpha, \beta \in \mathbb{R}$.

Of special interest to us is the case when $V = \mathbb{R}^n = \{(x = (x_1, x_2, \cdots, x_n) : x_j \in \mathbb{R}, j = 1, 2, \cdots, n\}$, the n-fold Cartesian product of $\mathbb{R}$ with itself. Addition is defined componentwise: $x + y = (x_1, x_2, \cdots, x_n) + (y_1, y_2, \cdots, y_n) \equiv (x_1 + y_1, x_2 + y_2, \cdots, x_n + y_n)$. The (scalar) product is $rx = (rx_2, rx_2, \cdots, rx_n)$ for each $r \in \mathbb{R}$ and $x \in V$. In one dimension the distance between $x, y \in \mathbb{R}$ is $d(x, y) = |x - y|$. There are many natural extensions of the absolute value function (to $\mathbb{R}^n$) that give us a notion of distance. A *norm* is such an extension. It is a function from $\mathbb{R}^n$ to $\mathbb{R}^+ = \{r \in \mathbb{R} : r \geq 0\}$ that assigns to each $x \in \mathbb{R}^n$ a non-negative real number $\|x\|$ in such a way that

(a) $\|x + y\| \leq \|x\| + \|y\|$ for all $x, y \in \mathbb{R}^n$,

(b) $\|\alpha x\| = |\alpha|\|x\|$ for $\alpha \in \mathbb{R}, x \in \mathbb{R}^n$,

(c) $\|x\| = 0$ iff $x = (x_1, x_2, \cdots, x_n) = (0, 0, \cdots, 0)$ is the zero vector.

Property (a) is known as Minkowski's inequality, (b) is referred to as positive homogeneity. From such a norm we obtain the "distance" between $x, y \in \mathbb{R}^n$ by letting $d(x, y) = \|x - y\|$.

More generally let us introduce a space S endowed with a notion of *distance* (also called a *metric*). Such a space, then, consists of a general set S and a function, d, defined as $S \times S$ into the non-negative reals $\mathbb{R}^+$. The basic property of d is the triangle inequality, $d(x, y) \leq d(x, z) + d(z, y)$, $d(x, y) = d(y, x)$, $d(x, y) = 0$ iff $x = y$, for $x, y, z \in S$. Such a space is called a *metric space* and (S, d) will denote it.

(HW) 1.30   Suppose d is defined, as above, by the norm $\|\cdot\|$. Prove the *triangle inequality*

$$d(x, y) \leq d(x, z) + d(z, y) \qquad \text{for } x, y, z \in \mathbb{R}^n.$$

Recall that there is a large class of norms on $\mathbb{R}^n$: Let $1 \leq p < \infty$ and $\|x\|_p = (\sum_{j=1}^{n} |x_j|^p)^{\frac{1}{p}}$. Also let $\|x\|_\infty := \max\{|x_1|, \cdots, |x_n|\}$ for any $x \in \mathbb{R}^n$. Properties (b) and (c) are clearly true. Minkowski's inequality is a consequence of another important inequality, *Hölder's inequality*.

Suppose $1 \leq p, q \leq \infty$ and $\dfrac{1}{p} + \dfrac{1}{q} = 1$ (when $q = \infty$, we let $\dfrac{1}{q} = 0$), then, letting $<$

$$x, y >:= \sum_{j=1}^{n} x_j y_j,$$

(*) $| < x, y > | \leq \|x\|_p \|y\|_p$

for all $x, y \in \mathbb{R}^n$. $< x, y >$ is called the *inner product* of x, y.

Let us see how (*) implies Minkowski's inequality. It is obvious that this last inequality is true for $p = 1$. Hence, suppose $1 < p < \infty$ and $\dfrac{1}{p} + \dfrac{1}{q} = 1$ (we say that in this case p and q are *conjugate indices*). The following equalities are equivalent:

$$(i) \quad \frac{1}{p} + \frac{1}{q} = 1, \qquad (ii) \quad \frac{p}{q} = p - 1, \qquad (iii) \quad (p-1)(q-1) = 1. \qquad (1.1)$$

Suppose $\quad x = (x_1, \cdots, x_n), y = (y_1, \cdots, y_n) \in \mathbb{R}^n \quad$ and $\quad u = (|x_1|, \cdots, |x_n|),$ $v = (|y_1|, \cdots, |y_n|), w = (|x_1 + y_1|^{p-1}, \cdots, |x_n + y_n|^{p-1})$, then

$$\|x + y\|_p^p = \sum_{j=1}^n |x_i + y_i||x_j + y_j|^{p-1} \leq | < u, w > | + | < v, w > |$$

$$= \sum_{j=1}^n |x_j||w_j| + \sum_{j=1}^n |y_j||w_j|$$

$$\leq \left(\sum_{j=1}^n |x_j|^p\right)^{\frac{1}{p}} \left(\sum_{j=1}^n |w_j|^q\right)^{\frac{1}{q}} + \left(\sum_{j=1}^n |y_j|^p\right)^{\frac{1}{p}} \left(\sum_{j=1}^n |w_j|^q\right)^{\frac{1}{q}}$$

$$= (\|x\|_p + \|y\|_p)\|w\|_q$$

(We used (*) to obtain the last inequality). On the other hand $\|w\|_q = \|x + y\|_p^{p-1}$ (use equality (ii) above). This last equality and the inequality before it immediately give us Minkowski's inequality.

As we stated in the introductory remarks, we assume you know elementary Calculus. We give a simple argument that gives us Hölder's inequality. Let φ be a continuous strictly increasing function that maps $\mathbb{R}^+$ into $\mathbb{R}^+$ such that $\varphi(0) = 0$.

Let ψ be the inverse function to φ: that is, $\psi(\varphi(x)) = x$ for all $x \geq 0$.

The shaded area A in Fig. 1.2 represents $\displaystyle\int_0^{x_0} \varphi(s)ds$, while the shaded area B repre-

sents $\displaystyle\int_0^{y_0} \psi(t)dt$. It is clear that the rectangle $[0, x_0] \times [0, y_0] = \{(x, y) \in \mathbb{R}^+ \times \mathbb{R}^+ : 0 \leq x \leq x_0, 0 \leq y \leq y_0\}$ has an area that is $\leq \displaystyle\int_0^{x_0} \varphi(s)ds + \int_0^{y_0} \psi(t)dt$:

$$x_0 y_0 \leq \int_0^{x_0} \varphi(s)ds + \int_0^{y_0} \psi(t)dt, \qquad (1.2)$$

with equality occuring iff $y_0 = \varphi(x_0)$.

Of interest to us are the cases when $\varphi(s) = s^{p-1}, p > 1$. In this case the inverse function to φ is $\psi(t) = t^{q-1}$ with $\dfrac{1}{p} + \dfrac{1}{q} = 1$; this is an immediate consequence of (1.1) (iii) (the equality

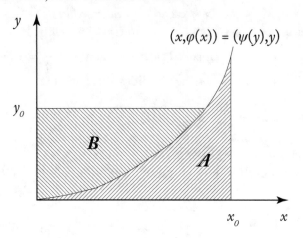

Figure 1.2: The graph of φ and ψ.

$(p-1)(q-1) = 1$). The inequality (1.2) is, then,

$$xy \leq \frac{x^p}{p} + \frac{y^q}{q} \tag{1.3}$$

for all $x, y \geq 0$ (just integrate $\varphi(s) = s^{p-1}$ and $\psi(t) = t^{p-1}$). To see how Hölder's inequality holds we first show that $|<u,v>| \leq 1$ when $u, v \in \mathbb{R}^n$ are vectors of norms 1: $\|u\|_p = 1 = \|v\|_q$: if $x, y \in \mathbb{R}^n$ are non-zero vectors, then $u = \dfrac{1}{\|x\|_p} x, v = \dfrac{1}{\|y\|_q} y$ and it follows that $|<u,v>| \leq \dfrac{1}{\|x\|_p} \dfrac{1}{\|y\|_q} |<x,y>|$ and the inequality $|<u,v>| \leq 1$ (if x or y is the zero vector, (1.3) is obvious). But, if u and v are unit vectors and p and q are conjugate indices,

$$|<u,v>| \leq \sum_{j=1}^{n} |u_j||v_j| \leq \sum_{j=1}^{n} \left(\frac{|u_j|^p}{p} + \frac{|v_j|^q}{q} \right)$$

$$= \frac{1}{p} \sum_{j=1}^{n} |u_j|^p + \frac{1}{q} \sum_{j=1}^{n} |v_j|^q = \frac{1}{p} + \frac{1}{q} = 1.$$

We realize that these notions (vector spaces, norms, distance functions) may be new to you. To make you feel better, think of the two dimensional case ($n = 2$), $p = 1, 2$ or ∞. In this case, you use $\| \ \|_1, \| \ \|_2, \| \ \|_\infty$ often (think about it).

(HW) 1.31   Draw the boundary of the unit disc in the plane ($\mathbb{R}^2$): $\{u = (x,y) \in \mathbb{R}^2 : (x^p + y^p)^{\frac{1}{p}} = \|u\|_p \leq 1\}$ for $1 \leq p \leq \infty$.

The special case $p = 2 = q$ is particularly important and simple. Play around with it. This is a special case of conjugate indices: $\dfrac{1}{2} + \dfrac{1}{2} = 1$. Hölder's inequality in this case is known as *Schwarz's inequality* (or *Cauchy-Schwarz inequality*). Convince yourself that the various arguments are really obvious.

1.5 CONTINUOUS FUNCTIONS, OPEN, CLOSED, AND COMPACT SETS IN $\mathbb{R}^n$

Functions mapping $S \subset \mathbb{R}^m$ into $\mathbb{R}^n$ can be (and often are) quite "wild". A condition that implies that a function is "reasonable" (as we shall see) for our purposes is *continuity*. We will define this concept for functions mapping a metric space V, with metric d^1, into a metric space W, with metric d^2 (both real vector spaces):

Definition 1.32   *Let $f : V = (V, d^1) \to W = (W, d^2)$. f is continuous at $x_0 \in V$ iff given any $\epsilon > 0$ there exists $\delta > 0$ such that if $x \in V$ and $d^1(x, x_0) < \delta$, then $d^2(f(x), f(x_0)) < \epsilon$. In this case we write $\lim_{x \to x_0} f(x) = f(x_0)$.*

Thus, V and W can be $\mathbb{R}^m$ and $\mathbb{R}^n$, each with one of the metrics defined in the last section.

We can reformulate this property in the following way. Usually we suppose $S \subset V$ is an *open subset*. By this we mean that if $x \in S$ then there exists an *open ball* $B = B_r(x) := \{y \in S : d(y, x) < r\}$ with center at x and radius $r > 0$ such that $B_r(x) \subset S$. The reformulation is that f is *continuous* at $x_0 \in S$ iff given an open ball in W with center at $f(x_0)$ then there exists an open ball with center at x_0, $B \subset S$, such that $f(B \cap S)$ is contained in the first ball. $f : S \to W$ is *continuous* on S iff it is continuous at each $x_0 \in S$.

The continuity of a function is determined by the open sets in the domain and the range. More precisely we have the following.

Theorem 1.33   *Suppose $f : S \to T$ is a function mapping a metric space (S, d) into a metric space (T, d'). f is continuous on S iff $f^{-1}(V) = \{s \in S : f(s) \in V\}$ is open in S whenever V is open in T.*

Comment. In general, whenever $f : S \to T$ (S and T general sets) and $V \subset T$, $f^{-1}(V) = \{s \in S : f(s) \in V\}$ is called the *inverse image of V*. The function f, thus, induces a mapping on the class of all subsets of T, 2^T, the power set of T, into the class of all subsets of S, 2^S, we denote this mapping by $f^{-1} : 2^T \to 2^S$. Whenever $\{V_\alpha\}$ is a collection of subsets of T,

$$f^{-1}\left(\bigcup_\alpha V_\alpha\right) = \bigcup_\alpha f^{-1}(V_\alpha), \quad f^{-1}\left(\bigcap_\alpha V_\alpha\right) = \bigcap_\alpha f^{-1}(V_\alpha). \tag{1.4}$$

(HW) 1.34   Prove these two equalities as well as $f^{-1}(V^c) = [f^{-1}(V)]^c$, where V^c is the complement of V in T.

Proof of Theorem 1.33: Suppose $f : S \to T$ is continuous and $V \subset T$ is open in (T, d'). We must show that $f^{-1}(V)$ is open in (S, d). Let $x \in f^{-1}(V)$. We shall show that there exists an open ball $B_\delta(x)$, centered at x and of radius $\delta > 0$, that lies in $f^{-1}(V)$. This will show that $f^{-1}(V)$ is open in (S, d). Since V is open in T, there exists $\epsilon > 0$ such that $B_\epsilon(f(x)) = \{y \in T : d'(y, f(x)) < \epsilon\} \subset V$. The continuity of f implies that there exists $\delta > 0$ such that if $d(y, x) < \delta$ then $d'(f(y), f(x)) < \epsilon$. That is, $f(y) \in B_\epsilon(f(x)) \subset V$. Hence $B_\delta(x) \subset f^{-1}(V)$ and it follows that $f^{-1}(V)$ is open.

Now suppose $f^{-1}(V)$ is open in S whenever V is open in T. Let $x \in S$ and $\epsilon > 0$. Consider $V = B_\epsilon(f(x))$ (you should verify that an open ball in a metric space is an open set). Then $f^{-1}(B_\epsilon(f(x)))$ is an open set in S and it contains x. Thus, there exists $\delta > 0$ such that $B_\delta(x) \subset f^{-1}(B_\epsilon(f(x)))$. But this assures us that $d'(f(y), f(x)) < \epsilon$ if $d(y, x) < \delta$ and we see that f is continuous at x. ∎

Remark. If (S, d) is a metric space so is (E, d) for any subset $E \subset S$.

(HW) 1.35   Use the notations in this "Remark". Show that $U \subset E$ is open in (E, d) iff there exists $V \subset S$, V open in (S, d), such that $U = V \cap E$. We see, therefore, that the notion of continuity applies to any function whose domain is a set $E \subset S$ and its range is in a metric space.

Definition 1.36   *Suppose $f : S \to T$ (S and T metric spaces). f is uniformly continuous iff given any $\epsilon > 0$, there exists $\delta > 0$ such that $d'(f(y), f(x)) < \epsilon$ whenever $d(x, y) < \delta$.*

Remark. Uniform continuity asserts that for each $\epsilon > 0$ we can find $\delta > 0$, independently of $x \in S$, such that $f(x)$ and $f(y)$ are less than ϵ apart as long as x and y are within δ of each other: the δ corresponds to ϵ, but is independent of $x \in S$.

There are many examples of continuous functions that are not uniformly continuous.

(HW) 1.37   Show that $f(x) = \sin \dfrac{1}{x}$, $f : S = (0, \infty) \to [-1, 1]$, is not uniformly continuous.

An important class of subsets of a metric space are the *compact sets*:

Definition 1.38   *A subset $C \subset S$ (a metric space) is compact iff whenever C is contained in the union of open subsets of S, then there is a finite collection of these sets whose union contains C.*

Simple examples of compact sets and non-compact sets are furnished by intervals in $\mathbb{R}$: $[a, b], -\infty < a < b < \infty$, is compact; however, $(a, b]$ is *not* compact. The sets $(a + \dfrac{1}{k}, b + 1)$, $k \in \mathbb{N}$, cover $(a, b]$, but no finite subcollection of these open intervals can cover this half-open interval (convince yourself of this fact).

We have encountered the open subsets of a metric space (S, d). A subset, F, of a metric space is *closed* iff its *complement* $U = F^c = \{y \in S : y \notin F\}$ is open. We shall find the properties that identify the closed sets and the compact sets.

The following result is almost immediate:

Theorem 1.39   *If C is a compact subset of $\mathbb{R}^n$, then C is bounded (i.e., it is contained in a ball).*

Proof: For each $x \in C$ consider the open ball $B_1(x)$. Then $C \subset \bigcup_{x \in C} B_1(x)$. Since C is compact there exists a finite number of points $\{x_1, x_2, \cdots, x_k\}$ in C such that $C \subset \bigcup_{j=1}^{k} B_1(x_j)$. Let $M = \max_{1 \le j, \ell \le k} d(x_j, x_\ell)$. We claim that $C \subset B_{M+1}(x_1)$. In fact, if $x \in C$, then $x \in B_1(x_j)$ for some $j = 1, 2, \cdots, k$. Thus, $d(x, x_1) \le d(x, x_j) + d(x_j, x_1) < 1 + M$. ∎

Let us now examine some properties of closed sets.

Theorem 1.40   *If $\{F_\alpha\}, \alpha \in \mathcal{A}$ is any collection of closed sets in $\mathbb{R}^n$, then*

(i) $\bigcap_{\alpha \in \mathcal{A}} F_\alpha$ *is closed, and*

(ii) if $F_1, F_2, \cdots, F_\ell$ is a finite collection of closed sets in $\mathbb{R}^n$, then $\bigcup_{j=1}^{\ell} F_j$ is closed.

(HW) 1.41   Show that if $U_\alpha, \alpha \in \mathcal{A}$, is any collection of open sets, then $\bigcup_{\alpha \in \mathcal{A}} U_\alpha$ is open and $\bigcap_{j=1}^{\ell} U_j$ is open, $\ell \in \mathbb{N}$. Use these facts to prove Theorem 1.40.

A characterization of closed sets in $\mathbb{R}^n$:

Theorem 1.42   *$F \subset \mathbb{R}^n$ is closed iff F contains all its limit points.*

Proof: Suppose F is closed and x_0 is a limit point of F that is not in F. Then $x_0 \in F^c = \mathbb{R}^n \setminus F = $ complement of F (in $\mathbb{R}^n$), which is, by definition, an open set. Thus, there exists $B_\epsilon(x_0) \subset F^c$ for some $\epsilon > 0$. But this means that x_0 is *not* a limit point of F (see the definition before Theorem 1.27). Thus, F must contain all its limit points.

Now suppose F contains all its limit points. Suppose $x_0 \in F^c$, then x_0 cannot be a limit point of F. Hence, there exists $\epsilon > 0$ such that $B_\epsilon(x_0)$ contains *no* point of F. This means that $B_\epsilon(x_0) \subset F^c$ and, thus, F^c is open. We conclude that F is closed. ∎

We can now establish

Theorem 1.43 $C \subset \mathbb{R}^n$ *is compact iff C is closed and bounded.*

Proof: Suppose C is compact. We already know it is bounded (Theorem 1.39). If C is not closed, then there is a limit point, x_0, of C that is not in C. Consider the sequence of open sets $\{U_k\}$, $k \in \mathbb{N}$, where $U_k = \{y \in \mathbb{R}^n : d(y, x_0) > \frac{1}{k}\}$. Clearly, $\bigcap_{k \geq 1} U_k^c = \{x_0\}$ and, thus, $\bigcup_{k \geq 1} U_k = \mathbb{R}^n \setminus \{x_0\} \supset C$. Hence, $\{U_k\}, k \geq 1$, is an open covering of C. Since C is compact, there are a finite number of these sets, $U_{k_1}, U_{k_2}, \cdots, U_{k_\ell}$ that cover C. If $k_j = \max\{k_1, k_2, \cdots, k_\ell\}$, then $C \subset \bigcup_{p=1}^{\ell} U_{k_p} \subset U_{k_j}$. But, then, $B_{\frac{1}{2k_j}}(x_0)$ is disjoint from C and, therefore, x_0 is not a limit point of C. This means that C is closed.

Now suppose $C(\subset \mathbb{R}^n)$ is closed and bounded. Let $\{U_\alpha : \alpha \in \mathcal{A}\}$ be an open covering of C. Since C is bounded, there exists $a > 0$ such that the n-fold Cartesian product $[-a, a] \times \cdots \times [-a, a]$ contains C. The set $V = C^c = \mathbb{R}^n \setminus C$ is open and, clearly, $\{U_\alpha : \alpha \in \mathcal{A}\} \cup \{V\}$ is an open covering of I.

Suppose we can show:

(*)   *Any open covering of a closed n-cube I has a finite subcovering of I.*

We then can conclude that there exists a finite subcovering $\{U_{\alpha_1}, \cdots, U_{\alpha_\ell}\}$ of $\{U_\alpha : \alpha \in \mathcal{A}\}$ that covers C. For we can apply (*) to the open covering $\{V\} \cup \{U_\alpha : \alpha \in \mathcal{A}\}$ of I to obtain the subcovering $\{U_{\alpha_1}, \cdots, U_{\alpha_\ell}\} \cup \{V\}$ of C (use the fact that V contains no point of C and, thus, $\{U_{\alpha_1}, \cdots, U_{\alpha_\ell}\}$ covers C).

Let us, then, give a proof of (*): Suppose I is n-fold Cartesian product of the interval $[-a, a]$, $a > 0$, and $\{U_\alpha : \alpha \in \mathcal{A}\}$ an open covering of I. Consider

$$S = \{y \in \mathbb{R} : 0 \leq y \leq a, \quad [-y, y] \times \cdots \times [-y, y] \ (n\text{-fold product})$$
$$\text{is covered by a finite number of the } U_\alpha's\}.$$

This set S of y's is not empty: $y = 0$ belongs to this set since $\{0, \cdots, 0\}$ is covered by one of the U_α's. By the *LUB* property of the reals, this bounded set of non-negative y's has a least upper bound y_0. If $y_0 = a$ we are done. If $y_0 < a$ there exists $\epsilon > 0$ such that $(y_0 - \epsilon, y_0 + \epsilon) \times \cdots \times (y_0 - \epsilon, y_0 + \epsilon) \subset U_{\alpha_0}$ an open set of our covering containing $(y_0, \cdots, y_0) \in \mathbb{R}^n$. But then y_0 is not the *LUB* we just described (since $(y_0 - \epsilon, y_0 + \epsilon)$ contains a point $y > y_0$ that is also in S). ∎

Theorem 1.43 is known as the *Heine-Borel Theorem.*

Theorem 1.44   Heine.   *Suppose $S \subset \mathbb{R}^m$ and $f : S \to \mathbb{R}^n$, where S is a compact subset and f is continuous. Then f is uniformly continuous.*

Proof: Suppose $m = 1 = n$. Let $\epsilon > 0$; then, for each $x \in S$ there exists $\delta = \delta(\epsilon, x)$ such that if $y \in S$ and $|y - x| < \delta$, then $|f(x) - f(y)| < \dfrac{\epsilon}{2}$. For each $x \in S$ let

$$I(x) = B_{\frac{\delta}{2}}(x) = (x - \frac{\delta}{2}, x + \frac{\delta}{2}),$$

the collection $\mathcal{U} = \{I(x) : x \in S\}$ is an open covering of S. By compactness $\mathcal{U}$ has a finite sub-covering of S. That is, there exists a finite set $\{x_1, x_2, \cdots, x_k\} \subset S$, such that

$$S \subset \bigcup_{\ell=1}^{k} I(x_\ell).$$

Let $\delta = \min\{\delta(\epsilon, x_\ell)/2 : \ell = 1, 2, \cdots, k\}$. Suppose $x, y \in S$ and $d(x, y) = |x - y| < \delta$. Then there exists $\ell \in \{1, 2, \cdots, k\}$ such that

$$x \in I(x_\ell) = B_{\delta(\epsilon, x_k)/2}(x_\ell).$$

Hence,

$$|y - x_\ell| \leq |y - x| + |x - x_\ell| < \delta + \frac{\delta(\epsilon, x_\ell)}{2} \leq \delta(\epsilon, x_\ell).$$

Therefore,

$$|f(y) - f(x)| \leq |f(y) - f(x_\ell)| + |f(x_\ell) - f(x)| < \frac{\epsilon}{2} + \frac{\epsilon}{2} = \epsilon.$$

■

(HW) 1.45   Write down a proof of Theorem 1.44 in the general case.

Theorem 1.46   *Suppose $S \subset \mathbb{R}^m$ is compact and $f : S \to \mathbb{R}^n$ is a continuous function. Then the range of f, $f(S)$, is a compact subset of $\mathbb{R}^n$.*

(HW) 1.47   Write out a proof of Theorem 1.46.

If T is a closed and bounded subset of $\mathbb{R}$, then T contains its greatest lower bound and its least upper bound (since each of these points are limit points of T, and T contains all its limit points by Theorem 1.42). A consequence of this fact and Theorem 1.46 is

Corollary 1.48   *Suppose f is a continuous real valued function defined on a compact domain, $\mathcal{D}(f)$, in $\mathbb{R}^n$, then f assumes its maximun and its minimum on $\mathcal{D}(f)$.*

More precisely, the range, $\mathcal{R}(f)$, of f being a bounded closed subset of $\mathbb{R}$, contains its *LUB* and *GLB*. Since these values are in the range of f, there exist $x, y \in \mathcal{D}(f) \ni f(x) = GLB\{f(s) : s \in \mathcal{D}(f)\}$ and $f(y) = LUB\{f(s) : s \in \mathcal{D}(f)\}$.

(HW) 1.49   Let $f : \mathbb{R}^2 \to \mathbb{R}$ be the function defined as follows:

$$f(x, y) = \begin{cases} \dfrac{xy}{x^2 + y^2} & \text{if } (x, y) \neq (0, 0) \\ 0 & \text{if } (x, y) = (0, 0). \end{cases}$$

Let $f_{x_0}(y) = f(x_0, y)$ for x_0 fixed and $f_{y_0}(x) = f(x, y_0)$. Show that f_{x_0} and f_{y_0} are continuous functions from $\mathbb{R}$ into $\mathbb{R}$; however, also show that f is *not* continuous at $(0, 0)$.

(HW) 1.50   Suppose $f : S \to T$ is a function mapping a metric space (S, d) into a metric space (T, d'). Show that f is continuous on S iff $f^{-1}(F)$ is closed in S for all F closed in T whenever F is closed in T (compare with Theorem 1.33).

If $f : S \to T$ we introduced the symbol f^{-1} to denote the function on the class of all subsets, V, of T defined by $f^{-1}(V) = \{s \in S : f(s) \in V\}$ (a subset of S). See Theorem 1.33 and the comment that follows (on page 19). If f is $1 : 1$ and onto T, then $f^{-1}(\{y\})$ is a set, $\{x\} \subset S$ for each $y \in T$. In this case, f^{-1} is also used to denote the function, whose domain is T and range is S, that maps y onto x. Clearly, $f^{-1}(f(x)) = x$ and $f(f^{-1}(y)) = y$ for all $x \in S$ and $y \in T$. f^{-1} is called, in this case, the *inverse function of f* .

(HW) 1.51   Suppose f is a $1 : 1$ continuous function on a compact domain $\mathcal{D}(f) \subset \mathbb{R}^m$ whose range, $\mathcal{R}(f)$, is in $\mathbb{R}^n$. Show that $f^{-1} : \mathcal{R}(f) \to \mathcal{D}(f)$ is continuous (Hint : use (H,W)(1.5.10)).

(HW) 1.52   Show that a subset C of $\mathbb{R}^n$ is compact iff each sequence in C has a convergent subsequence with limit in C.

CHAPTER 2

Differentiation on $\mathbb{R}^n$

2.1 DIFFERENTIABILITY ON $\mathbb{R}^n$

We assume you know that a function defined on an open interval $(a, b) \subset \mathbb{R}$ that is real valued is *differentiable* at $x_0 \in (a, b)$ iff

$$\lim_{\delta \to 0,\, \delta \neq 0} \frac{f(x_0 + \delta) - f(x_0)}{\delta} = f'(x_0) \tag{2.1}$$

exists (we assume δ to be sufficiently small so that $x_0 + \delta \in (a, b)$). The limit $f'(x_0)$, when it exists, is called the *derivative of f at x_0*. $f'(x_0)$ is the slope of the line that is tangent to the graph of f at the point $(x_0, f(x_0))$ (see Fig. 2.1).

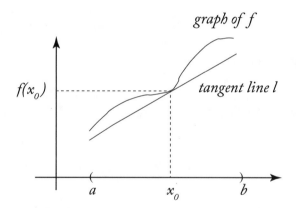

Figure 2.1: The graphs of f and ℓ.

This line has the form

$$y = \ell(x) = f(x_0) + f'(x_0)(x - x_0) \tag{2.2}$$

for all $x \in \mathbb{R}$. Any function ℓ that satisfies $y = \ell(x) = \alpha x + \beta$ for all $x \in \mathbb{R}$ has a graph that is a straight line with slope α that intersects the y-axis in the point $(0, \beta)$. The ℓ satisfying (2.2) is the "best" linear approximation to f at the point $(x_0, f(x_0))$; we have

$$f(x_0 + t) - f(x_0) = f'(x_0)t + \epsilon(t) \tag{2.3}$$

when $\dfrac{\epsilon(t)}{t} \to 0$ as $t \to 0$ ($t \neq 0$). You should convince yourself that (2.3) together with the condition on the "error term" $\epsilon(t)$ (that $\displaystyle\lim_{t\to 0, t\neq 0} \dfrac{\epsilon(t)}{t} = 0$) is equivalent to the existence of the derivative $f'(x_0)$.

We will extend the notion of a derivative to higher dimensions based on the property presented by (2.3). In order to do this we shall give the shortest course on "linear transformations" mapping $\mathbb{R}^m$ into $\mathbb{R}^n$ that has ever been given to humans. Recall that $\mathbb{R}^n, n \in \mathbb{N}$, is a vector space.

Definition 2.1   *A function $T : \mathbb{R}^m \to \mathbb{R}^n$ is a linear transformation iff*

$$T(\alpha x + \beta y) = \alpha T(x) + \beta T(y)$$

for all $\alpha, \beta \in \mathbb{R}$ and $x, y \in \mathbb{R}^m$.

(HW) 2.2   Show that when $m = 1 = n$ a linear transformation is a function whose graph is a straight line through the origin with slope $r \in \mathbb{R}$.

Definition 2.3   *A finite set of vectors $\{x_1, x_2, \cdots, x_\ell\}$ in $\mathbb{R}^n$ is linearly independent iff $\alpha_1 x_1 + \cdots + \alpha_\ell x_\ell = 0$ for $\alpha_j \in \mathbb{R}$, $j = 1, 2, \cdots, \ell$, implies $\alpha_j = 0$, $j = 1, 2, \cdots, \ell$. If such a set is not linearly independent it is said to be linearly dependent.*

Let $e_j = (0, \cdots, 0, \overset{\text{↕ } j-th \ place}{1}, 0, \cdots, 0) \in \mathbb{R}^n$, $j = 1, 2, \cdots, n$. These vectors are known as the *canonical basis vectors* of $\mathbb{R}^n$. Clearly, each $v = (\alpha_1, \cdots, \alpha_n) \in \mathbb{R}^n$ is a linear combination of these canonical basis vectors:

$$(*) \qquad\qquad v = \sum_{j=1}^{n} \alpha_j e_j.$$

More generally by a *linear combination* of a finite set of vectors $\{f_1, f_2, \cdots, f_\ell\}$ in $\mathbb{R}^n$ we mean a vector x of the form $x = \displaystyle\sum_{j=1}^{\ell} \alpha_j f_j$, $\alpha_j \in \mathbb{R}^n$.

(HW) 2.4   Show that the vectors $e_1, e_2, \cdots, e_n$ are linearly independent (Hint: use the fact that they are *orthonormal*; that is, $< e_i, e_j >= \delta_{ij}$, where

$$\delta_{ij} = \begin{cases} 0 & \text{if} \quad i \neq j \\ 1 & \text{if} \quad i = j \end{cases}$$

is the *Kronecker δ-function*).

Definition 2.5   *If $\{v^{(j)}; j \in \mathbb{N}\}$ is a collection of vectors, the span of $\{v^{(j)}, j \in \mathbb{N}\}$ is the set of all finite linear combinations,* $\displaystyle\sum_{finite} \alpha_j v^{(j)}$, $\alpha_j \in \mathbb{R}$.

You should check that any such span is a vector space over $\mathbb{R}$.

We see, therefore, that $\mathbb{R}^n$ is spanned by $\{e_1, e_2, \cdots, e_n\}$ (see (*) above).

Definition 2.6   *If $b_1, b_2, \cdots, b_\ell$ is a linearly independent set of vectors in $\mathbb{R}^n$ whose span is a linear subspace V of $\mathbb{R}^n$, this set is called a basis of V.*

A fundamental result in linear algebra (the study of vector spaces) is that, if $\{b_1, b_2, \cdots, b_\ell\}$ is a basis of V, any other basis has the same number, ℓ, of members. This number is called the *dimension* of the vector space V.

Each of you probably considered $\mathbb{R}^n$ to be of "dimension n." The fact that the canonical basis vectors is a set of n vectors shows, by this last definition, that $\mathbb{R}^n$ has, indeed, dimension n (We have tacitly assumed that you realized that $\{e_1, e_2, \cdots, e_n\}$ is a basis of $\mathbb{R}^n$).

(HW) 2.7   Find two other bases $\{f_1, f_2, f_3\} \equiv \mathcal{F}$, $\{g_1, g_2, g_3\} \equiv \mathcal{G}$ of $\mathbb{R}^3$ (besides $\{e_1, e_2, e_3\}$) such that $\mathcal{F}$ is *orthonormal* ($< f_j, f_i >= \delta_{ji}$, $i, j = 1, 2, 3$) and $\mathcal{G}$ is not orthonormal.

We will now examine how to express linear transformations $T : \mathbb{R}^m \to \mathbb{R}^n$ in terms of bases of $\mathbb{R}^m$ and $\mathbb{R}^n$. Let $\{e_1^{(m)}, \cdots, e_m^{(m)}\}$ and $\{e_1^{(n)}, \cdots, e_n^{(n)}\}$ be canonical bases of these two spaces. Since $T e_j^{(m)} \in \mathbb{R}^n$, there exist real numbers $t_{1j}, t_{2j}, \cdots, t_{nj}$ such that $\displaystyle T e_j^{(m)} = \sum_{i=1}^{n} t_{ij} e_i^{(n)}$. Thus, if $\displaystyle x = \sum_{j=1}^{m} a_j e_j^{(m)}$ is a vector in $\mathbb{R}^m$ ($a_j \in \mathbb{R}$ for $j = 1, \cdots, m$), we have, by the linearity of T,

$$Tx = \sum_{j=1}^{m} a_j T e_j^{(m)} = \sum_{j=1}^{m} a_j \sum_{i=1}^{n} t_{ij} e_i^{(n)} = \sum_{i=1}^{n} \left(\sum_{j=1}^{m} a_j t_{ij}\right) e_i^{(n)}.$$

That is, if $x \in \mathbb{R}^m$ is the vector having coefficients $a_1, a_2, \cdots, a_m$ in its expansion in terms of the canonical basis $\{e_j^{(m)}, j = 1, \cdots, m\}$ of $\mathbb{R}^m$, then Tx is the vector in $\mathbb{R}^n$ whose expansion in terms of the 2nd canonical basis has coefficients

$$b_i = \sum_{j=1}^{m} a_j t_{ij}, \; i = 1, 2, \cdots, n.$$

We can write this equality in "matrix notation":

$$
\begin{bmatrix} b_1 \\ b_2 \\ \vdots \\ b_n \end{bmatrix} = \begin{bmatrix} t_{11} & t_{12} & \cdots & t_{1m} \\ t_{21} & t_{22} & \cdots & t_{2m} \\ \cdot & \cdot & \cdots & \cdot \\ t_{n1} & t_{n2} & \cdots & t_{nm} \end{bmatrix} \begin{bmatrix} a_1 \\ a_2 \\ \vdots \\ a_m \end{bmatrix}. \tag{2.4}
$$

That is, we write the coefficients of x in a column and (2.4) produces the coefficients of Tx written, again, as a column. You can consider (2.4) as the formula that produces the coefficients of Tx written, again, as a column. You can consider (2.4) as the formula that produces

$$
b = \begin{bmatrix} b_1 \\ \vdots \\ b_n \end{bmatrix} \text{ as a "product" of the matrix } \begin{bmatrix} t_{11} & t_{12} & \cdots & t_{1m} \\ t_{21} & t_{22} & \cdots & t_{2m} \\ \cdot & \cdot & \cdots & \cdot \\ t_{n1} & t_{n2} & \cdots & t_{nm} \end{bmatrix} = (t_{ij}) \equiv [T] \text{ and the column}
$$

$$
\text{vector } \begin{bmatrix} a_1 \\ a_2 \\ \vdots \\ a_n \end{bmatrix} = a:
$$

$$
b = [T]a = (t_{ij})a. \tag{2.5}
$$

Convince yourself that the argument we gave is independent of the basis we chose in $\mathbb{R}^m$ and the one we chose in $\mathbb{R}^n$; we only used the uniqueness of the representation of a vector with respect to a basis.

(HW) 2.8   Suppose $S : \mathbb{R}^m \to \mathbb{R}^n$ and $T : \mathbb{R}^n \to \mathbb{R}^p$ are linear transformations and their matrices (as described above) are

$$
[S] = \begin{bmatrix} s_{11} & s_{12} & \cdots & s_{1m} \\ s_{21} & s_{22} & \cdots & s_{2m} \\ \cdot & \cdot & \cdots & \cdot \\ s_{n1} & s_{n2} & \cdots & s_{nm} \end{bmatrix}
$$

and

$$
[T] = \begin{bmatrix} t_{11} & t_{12} & \cdots & t_{1n} \\ t_{21} & t_{22} & \cdots & t_{2n} \\ \cdot & \cdot & \cdots & \cdot \\ t_{p1} & t_{p2} & \cdots & t_{pn} \end{bmatrix}.
$$

Consider the composite mapping $P = TS : \mathbb{R}^m \to \mathbb{R}^p$ $((TS)(x) = T(S(x)))$. Show that its matrix satisfies

$$
(p_{ij}) = [P] = [TS] = [T][S], \tag{2.6}
$$

where $p_{ij} = \sum_{\ell=1}^{n} t_{i\ell}s_{\ell j}$, $j = 1, 2, \cdots, m$, $i = 1, 2, \cdots, p$. This shows that the composition of T and S (which is a "product") has a matrix that is the "product" of the matrix of T times the matrix of S.

Let us go back to the notion of differentiability we introduced at the beginning of Chapter 2.

Definition 2.9 *A function f defined on an open ball about $x_0 \in \mathbb{R}^m$ having values in $\mathbb{R}^n$ is differentiable at x_0 iff there exists a linear transformation $T : \mathbb{R}^m \to \mathbb{R}^n$ such that*

$$\lim_{h \to 0 \in \mathbb{R}^m, h \neq 0} \frac{f(x_0 + h) - (f(x_0) + Th)}{\|h\|} = 0. \tag{2.7}$$

The linear transformation T is called the *derivative of f at x_0* (sometimes it is called the *differential of f at x_0*) and is denoted by $(Df)(x_0)$ or $df(x_0)$. It is easy to see that T is unique.

(HW) 2.10 Let $T : \mathbb{R}^m \to \mathbb{R}^n$ be a linear transformation and $[T] = (t_{ij})$ its matrix. Show that there exists $C > 0$

$$\|Tx\| \leq C\|x\| \tag{2.8}$$

for all $x \in \mathbb{R}^m$ (the first norm is the $\mathbb{R}^n$ norm and the second norm is the $\mathbb{R}^m$ norm). Hint: Let

$$\||T\|| = (\sum_{\substack{1 \leq i \leq n \\ 1 \leq j \leq m}} |t_{ij}|^2)^{\frac{1}{2}}$$

and use (2.4) and Schwarz's inequality to show that $c \leq \||T\||$).

Definition 2.11 *The inf$\{C > 0 : (2.8)$ holds for all $x \in \mathbb{R}^m\}$ is called the norm of T and is denoted by $\|T\|$. $\||T\||$ is called the Hilbert–Schmidt norm of T.*

(HW) 2.12 Use (2.8) to show that any linear transformation is a uniformly continuous function.

(HW) 2.13 Show that if f is differentiable at x_0, then it is continuous at x_0.

We see from this last exercise that differentiability is a stronger condition than continuity. As we progress, we shall see in what ways differentiability implies "more smoothness" than continuity.

When a student is first introduced to the notion of a derivative of a function of several variables she often encounters "directional derivatives". By a *direction* in $\mathbb{R}^m$ we mean a unit vector θ. Suppose f is a function whose domain includes an open ball of radius $r > 0$ about $x_0 \in \mathbb{R}^m$ and whose range lies in $\mathbb{R}^n$.

Definition 2.14   *For a function f, as described in the last statement, and a direction θ, form the function $\varphi(t) = f(x_0 + t\theta)$ for $t \in (-r, r)$. If φ is differentiable at 0, then $\varphi'(0)$ is called the directional derivative of f at x_0 in the direction θ. We denote this value by $D_\theta f(x_0)$.*

It is clear that this is equivalent to

$$(D_\theta f)(x_0) = \lim_{t \to 0, t \neq 0} \frac{1}{t}(f(x_0 + t\theta) - f(x_0)) \tag{2.9}$$

(realize that $f(x_0 + t\theta) - f(x_0) \in \mathbb{R}^n$; thus, this is a bit more general than the one dimensional definition of a derivative). If $\theta = e_j$, $1 \le j \le m$, one of the elements of the canonical basis for $\mathbb{R}^n$, $(D_{e_j} f)(x_0)$ is called the *partial derivative of f with respect to x_j ($x = (x_1, \cdots, x_j, \cdots, x_m) \in \mathbb{R}^m$) at x_0*. Common notation for this value is $D_j f(x_0)$, or $\frac{\partial f}{\partial x_j}(x_0)$, or $f_{x_j}(x_0)$.

(HW) 2.15   Show that if $x_0 = (x_1^0, x_2^0, \cdots, x_m^0)$, then

$$(D_j f(x_0) =$$
$$\lim_{x_j \to x_j^0, x_j \neq x_j^0} \frac{f(x_1^0, \cdots, x_{j-1}^0, x_j, x_{j+1}^0 \cdots, x_m^0) - f(x_1^0, \cdots, x_{j-1}^0, x_j^0, x_{j+1}^0, \cdots, x_m^0)}{x_j - x_j^0}. \tag{2.10}$$

We will see that the partial derivatives (and, more generally, the directional derivatives) are very useful for calculations and (2.10) is important for this; however, their existence (without more assumptions) does not say very much about the function.

The following exercises illustrate this last statement:

(HW) 2.16   Let $f(x, y) = (xy)^{\frac{1}{3}}$ for all $(x, y) \in \mathbb{R}^2$ (why is f defined for all (x, y)?). Show that both partial derivatives of f exist at $(0, 0)$ and equal 0.

Formula (2.10) tells us that the partial derivative D_j is really one-dimensional. We fix the variables $x_1, \cdots, x_{j-1}, x_{j+1}, \cdots, x_m$ to their values at x_0 and let the jth variable vary about x_j^0. We obtain, therefore, a function of one variable and $(D_j f)(x_0)$ is the derivative at x_j^0 of this function.

Since we can calculate the derivative of many real valued functions of one variable, we know that a large class of partial derivatives have simple expressions. We use this and further assumptions on these partial derivatives to derive many properties of differentiable functions.

Theorem 2.17   *If f is differentiable at x_0, then f has a derivative in every direction at x_0 and*

$$(D_\theta f)(x_0) = df(x_0)\theta \tag{2.11}$$

(HW) 2.18   Prove Theorem 2.17.

In particular, if f is differentiable at x_0 each of the partial derivatives $\dfrac{\partial f}{\partial x_j}(x_0)$, $j = 1, \cdots, m$, exist. An important (but easily provable) result is the following formula for the matrix of $Df(x_0)$ with respect to the canonical bases of $\mathbb{R}^m$ and $\mathbb{R}^n$:

Theorem 2.19   *If $f : \mathcal{D}(f) \subset \mathbb{R}^m \to \mathbb{R}^n$ is differentiable at x_0 then the matrix of $Df(x_0)$ (with respect to the two canonical bases) is*

$$[Df(x_0)] = \begin{bmatrix} \dfrac{\partial f_1}{\partial x_1} & \cdots & \dfrac{\partial f_1}{\partial x_m} \\ \vdots & & \vdots \\ \dfrac{\partial f_n}{\partial x_1} & \cdots & \dfrac{\partial f_n}{\partial x_m} \end{bmatrix}, \tag{2.12}$$

where each partial derivative is evaluated at x_0 and $f = (f_1, f_2, \cdots, f_n) \in \mathbb{R}^n$.

The proof of this theorem is really rather simple; nevertheless, if this is the first time you have been exposed to these vector space notions, you might be a bit overwhelmed. Let us look at the simpler situation when $n = 1$ (that is, f is real-valued). In this case the matrix in (2.4) is $1 \times m$ (1 row and m columns); it is a "row vector" $(t_{11}, t_{12}, \cdots, t_{1m}) = (t_1, t_2, \cdots, t_m)$ (forget about the first index, which is 1). Thus, (2.4) has the form

$$(*) \qquad\qquad b = (t_1, t_2, \cdots, t_m) \begin{pmatrix} a_1 \\ a_2 \\ \vdots \\ a_m \end{pmatrix}.$$

Convince yourself that the right side of $(*)$ can be interpreted as the inner product of $x = t_1 e_1 + \cdots + t_m e_m$ and $y = a_1 e_1 + \cdots + a_m e_m = <x, y>$. An easy calculation is that in this case $[(Df(x_0)]$ is the *gradient* of f evaluated at x_0:

$$(\nabla f)(x_0) \equiv [(Df)(x_0)] = (\frac{\partial f}{\partial x_1}(x_0), \frac{\partial f}{\partial x_2}(x_0), \cdots, \frac{\partial f}{\partial x_m}(x_0)). \tag{2.13}$$

This is equality (2.12) in this case (in which $f(x) \in \mathbb{R} = \mathbb{R}^1$). The general case follows from this one by applying the argument we just gave to each component, f_j, of $f = (f_1, f_2, \cdots, f_m)$, $j = 1, 2, \cdots, n$. Because the gradient is so important we repeat its definition:

Definition 2.20   *The vector on the right of equality (2.13) is called the gradient of the real valued function f. One usually uses the notation $(\nabla f)(x_0)$ for the gradient of f at x_0. Consider the gradient to be the operator*

$$\nabla = (\frac{\partial}{\partial x_1}, \frac{\partial}{\partial x_2}, \cdots, \frac{\partial}{\partial x_m}) \text{ applied to the scalar valued function } f,$$

$$\nabla f = (\frac{\partial}{\partial x_1}, \frac{\partial}{\partial x_2}, \cdots, \frac{\partial}{\partial x_m}) f = (\frac{\partial f}{\partial x_1}, \frac{\partial f}{\partial x_2}, \cdots, \frac{\partial f}{\partial x_m})$$

(evaluated at x_0 in order to obtain (2.13)).

The matrix in (2.12) is called the *Jacobian matrix of f* (having values in $\mathbb{R}^n$).

(HW) 2.21   Suppose $f = (f_1, f_2, \cdots, f_n) \in \mathbb{R}^n$ is differentiable about x_0 in the domain of $f = D(f)$. Show that each of the component functions $f_1, f_2, \cdots, f_n$ is differentiable at x_0. The converse is also true.

We have already presented good evidence that differentiability is an important property and that the partial derivative, whose existence, by itself, does not imply much, is often easy to calculate. We shall now show how to determine the differentiability of a function by using the partial derivatives:

Theorem 2.22   *Suppose $f : D(f) \subset \mathbb{R}^m \to \mathbb{R}^n$ and each of the partial derivatives of f exist in an open ball with center x_0 in $D(f)$ and are continuous in this ball, then f is differentiable at x_0.*

We will make use of the *Mean Value Theorem* which has the following general statement:

Theorem 2.23   *Let f and g be continuous real valued functions on the finite closed interval $I = [a, b]$ that are differentiable at each point interior to I, then there exists a point $\xi \in (a, b)$ such that*

$$[f(b) - f(a)] g'(\xi) = [g(b) - g(a)] f'(\xi). \tag{2.14}$$

Comment: The usual statement of this result is the special case when $g(x) \equiv x$: there exists $\xi \in (a, b)$ such that

$$f(b) - f(a) = (b - a) f'(\xi). \tag{2.14'}$$

We will discuss the easy proof of (2.14) in a short while. Let us accept it for the moment and use it for the proof of Theorem 2.22. We reduced this result to the case $n = 1$. We will now prove it when $m = 2$ (it will then be obvious to extend it to the general case).

Let $f : \mathcal{D}(f) \to \mathbb{R}^1$, where the domain of f, $\mathcal{D}(f)$, is an open subset of $\mathbb{R}^2$. If $x_0 \in \mathcal{D}(f)$ let $\epsilon > 0$ be such that $\overline{B_\epsilon(x_0)} \subset \mathcal{D}(f)$. The partial derivatives $(D_j f)(x) = \dfrac{\partial f}{\partial x_j}(x)$, $j = 1, 2$, are uniformly continuous in $\overline{B_\epsilon(x_0)}$. Let T be the linear transformation on $\mathbb{R}^2$ into $\mathbb{R}^1$ mapping $h = (h_1, h_2)$ into

$$Th = (\nabla f)(x_0) \cdot h = ((D_1 f)(x_0), (D_2 f)(x_0)) \cdot (h_1, h_2).$$

Suppose $\|h\| < \epsilon$. We then have, letting $x_0 = (x_1^0, x_2^0)$

$$
\begin{aligned}
&f(x_0 + h) - f(x_0) \\
&= \{f(x_1^0 + h_1, x_2^0 + h_2) - f(x_1^0, x_2^0 + h_2)\} + \{f(x_1^0, x_2^0 + h_2) - f(x_1^0, x_2^0)\} \\
&= (D_1 f)(\xi, x_2^0 + h_2) h_1 + (D_1 f)(x_1^0, \eta) h_2,
\end{aligned}
$$

where ξ lies between $x_1^0 + h_1$ and x_1^0 and η lies between $x_2^0 + h_2$ and x_2^0 (we are using (2.14'))
Let $s = (\xi, x_2^0 + h_2)$ and $t = (x_1^0, \eta)$. We have shown

$$f(x_0 + h) - f(x_0) = (D_1 f)(s)h_1 + (D_2 f)(t)h_2,$$

where $s, t \in B_\epsilon(x_0)$. Observe, also, that $s, t \to x_0$ as $h \to 0$. Consequently,

$$
\begin{aligned}
&f(x_0 + h) - f(x_0) - (\nabla f)(x_0) \cdot h \\
&= ((D_1 f)(s) - (D_1 f)(x_0), (D_2 f)(t) - (D_2 f)(x_0)) \cdot h \\
&:= e(h).
\end{aligned}
$$

Using Schwarz's inequality we have

$$|e(h)| \leq [\,|(D_1 f)(s) - (D_1 f)(x_0)|^2 + |(D_2 f)(t) - (D_2 f)(x_0)|^2]^{\frac{1}{2}} \|h\|$$

Observe that $s, t \to x_0$ as $h \to 0 \in \mathbb{R}^2$. Thus, by the continuity of $D_j f$, $j = 1, 2$, we see that the term $[\cdots]^{\frac{1}{2}} \to 0$ as $h \to 0$. We have proved, therefore, that

$$(\dagger) \qquad \lim_{h \to 0,\, h \neq 0} \frac{f(x_0 + h) - f(x_0) - (\nabla f)(x_0) \cdot h}{\|h\|} = 0.$$

The equality (2.7) with the linear transformation T is represented by the 1×2 matrix $(\nabla f)(x_0) =$ gradient of f evaluated at x_0. If you compare with equality (2.12) this is precisely $(Df)(x_0)$ when $m = 2$ and $n = 1$. Hence, $(\dagger)$ tells us that f is differentiable at x_0, which is what we wanted to prove. It should be fairly clear that the case $m > 2$ follows by an easy extension of this argument. ∎

We promised to establish the Mean Value Theorem, Theorem 2.23. We will do this with your help. Recall that we had the general form (2.14) and the special form (2.14').

(HW) 2.24   Prove the Mean Value Theorem in the 2.14 setting. (Hint: Let $h(x) = [f(b) - f(a)] g(x) - [g(b) - g(a)] f(x)$ for $x \in [a, b]$. Observe that $h(a) = h(b)$ and deduce that there exists $\xi \in (a, b)$ such that $h'(\xi) = 0$).

If $f : \mathcal{D}(f) \to \mathcal{R}(f)$, $g : \mathcal{D}(g) \to \mathcal{R}(g)$ and $\mathcal{D}(g) \supset \mathcal{R}(f)$ ($\mathcal{D}(h)$ denotes the domain of a function h and $\mathcal{R}(h)$ its range), then we can define the *composite* $g \circ f : \mathcal{D}(f) \to \mathcal{R}(g)$ to be the function whose value $g \circ f(x)$ at $x \in \mathcal{D}(f)$ is $g(f(x)) \in \mathcal{R}(g)$:

$$g \circ f = g(f(x)), \quad \text{for all } x \in \mathcal{D}(f).$$

As is the case when considering one variable real valued functions, the differentiation of composite functions can be expressed in terms of the derivatives of the factors in the higher dimensional case. We have

Theorem 2.25   *Suppose $f : \mathcal{D}(f) \subset \mathbb{R}^\ell \to \mathbb{R}^m$ is differentiable at x_0 and $g : \mathcal{D}(g) \subset \mathbb{R}^m \to \mathbb{R}^n$, with $\mathcal{R}(f) \subset \mathcal{D}(g)$, is differentiable at $f(x_0)$, then $g \circ f$ is differentiable at x_0 and*

$$d(g \circ f)(x_0 = dg(f(x_0)) \circ df(x_0)^1$$

Proof: Our hypothesis tell us that there exists a linear map $S : \mathbb{R}^\ell \to \mathbb{R}^m$ and linear map $T : \mathbb{R}^m \to \mathbb{R}^n$ such that

(†) $$f(x_0 + h) = f(x_0) + Sh + \|h\|\epsilon(h) \text{ with } \epsilon(h) \to 0 \text{ as } h \to 0$$

and $g(y_0 + k) = g(y_0) + Tk + \|k\|\eta(k)$ with $\eta(k) \to 0$ as $k \to 0$, where $y_0 = f(x_0)$.

Define k by letting $f(x_0) + k = f(x_0 + h)$. Then

$$\begin{aligned}
&g \circ f(x_0 + h) - g \circ f(x_0) \\
&= g(f(x_0) + k) - g(f(x_0)) \\
&= Tk + \|k\|\eta(k) \\
&= T(f(x_0 + h) - f(x_0)) + \|k\|\eta(k) + T(Sh + \|h\|\epsilon(h)) + \|k\|\eta(k) \\
&= T \circ Sh + \|h\|T\epsilon(h) + \|k\|\eta(k) \\
&= T \circ Sh + \|h\|(T\epsilon(h) + \frac{\|k\|}{\|h\|}\eta(k)).
\end{aligned}$$

Thus, we have to show that $T\epsilon(h) + \dfrac{\|k\|}{\|h\|}\eta(k) \to 0$ as $h \to 0, h \neq 0$. But $\|T\epsilon(h)\| \leq \|T\|\|\epsilon(h)\|$, where $\|T\|$ is the norm of the linear transformation T (see page 29) and the other two norms are the Euclidean norm in $\mathbb{R}^m$ and the Euclidean norm in $\mathbb{R}^n$. Since $\epsilon(h) \to 0$ as $h \to 0$, all we need to show is that

(∗) $$\lim_{h \to 0, h \neq 0} \frac{\|k\|}{\|h\|}\eta(k) = 0.$$

---

[1] Remember: $df(x_0)$ is a linear transformation from $\mathbb{R}^\ell$ into $\mathbb{R}^m$ and $dg(f(x_0))$ is a linear transformation from $\mathbb{R}^m$ into $\mathbb{R}^n$; thus, the composition $dg(f(x_0)) \circ df(x_0)$ is well defined.

But, by (†) and the definition of k, we see that $k = Sh + \|h\|\epsilon(h)$. Thus, $\|k\| \leq \|h\|(\|S\| + \|\epsilon(h)\|)$. First note that this implies that $\|k\| \to 0$ as $\|h\| \to 0$ and, secondly, for $h \neq 0$, $\dfrac{\|k\|}{\|h\|}\|\eta(k)\| \leq (\|S\| + \|\epsilon(h)\|)\|\eta(k)\| \to 0$ as $h \to 0$. $\blacksquare$

(HW) 2.26   Under the same hypotheses of Theorem 2.25 show that

$$\frac{\partial}{\partial x_j}(g \circ f)_i(x_0) = \sum_{k=1}^{m} \frac{\partial g_i}{\partial y_k}(f(x_0))\frac{\partial f_k}{\partial x_j}(x_0), \tag{2.15}$$

where $(g \circ f)_i$, g_i are the ith components of $g \circ f$ and g, $i = 1, 2, \cdots, n$, f_k is the kth component of f, $k = 1, 2, \cdots, m$ and $g(y) = (g_1(y), \cdots, g_n(y))$ for $y = (y_1, \cdots, y_m) \in \mathbb{R}^m$.

Equality (2.15) is, notationally, somewhat of a nightmare. The following equality (not to be taken too seriously) serves as a help for remembering formula (2.15): Let $y = f(x)$ and $z = g(y)$, then

$$\frac{\partial z_i}{\partial x_j} = \sum_{k=1}^{n} \frac{\partial z_i}{\partial y_k}\frac{\partial y_k}{\partial x_j}. \tag{2.15'}$$

$\dfrac{\partial z_i}{\partial y_k}$ and $\dfrac{\partial y_k}{\partial x_k}$ should *not* be multiplied together and the ∂y_k's cancelled!

Of course, (2.15) does provide us a formula for the partial derivatives of a composite function.

(HW) 2.27   Suppose we have the same hypotheses as in Theorem 2.25 except that $\ell = 1 = n$. Show that if f is differentiable at x_0 and g is differentiable at $f(x_0)$, then $(g \circ f)' = <\nabla g, f'>$, where $(g \circ f)'$ is evaluated at x_0 ($\in \mathbb{R}$) (and so is f') and ∇g is evaluated at $f(x_0)$.

2.2   HIGHER PARTIAL DERIVATIVES AND TAYLOR'S THEOREM

In one dimension suppose $I = (a, b)$ is an open interval in $\mathbb{R}$ and $f : I \to \mathbb{R}$. We will now consider stronger notions of smoothness of f on I: We say f belongs to the *class $C^n(I)$*, $n \geq 1$, if f has derivatives $f^{(k)}$ of all orders k, $1 \leq k \leq n$, and $f^{(n)} = \dfrac{d^n}{dx^n}f$ is continuous on I. If $n = 0$, $f \in C^0(I)$ just means continuity on I. We would like to extend this notion to higher dimensions. In order to do this we need to develop some facts about higher derivatives in dimension $m > 1$.

Suppose B is an open ball in $\mathbb{R}^m$ that is not empty. Let us understand some of the properties of the higher partial derivatives of f. That is, suppose $D_1 f, \cdots, D_m f = \dfrac{\partial f}{\partial x_1}, \cdots, \dfrac{\partial f}{\partial x_m}$ are each well defined on B for $f : B \to \mathbb{R}^1$. We may wonder when each of these m functions have partial derivatives on B. We will see that the continuity of these partial derivatives plays an important role (as was the case in Theorem 2.22). Consider the following function defined on $\mathbb{R}^2$:

$$f(x, y) = \begin{cases} xy \dfrac{x^2 - y^2}{x^2 + y^2} & \text{if } (x, y) \neq (0, 0) \\ 0 & \text{if } (x, y) = (0, 0). \end{cases}$$

(HW) 2.28   Show the existence of all the second partial derivatives

$$\frac{\partial}{\partial x} \frac{\partial f}{\partial y}, \frac{\partial}{\partial x} \frac{\partial f}{\partial x}, \frac{\partial}{\partial y} \frac{\partial f}{\partial x} \text{ and } \frac{\partial}{\partial y} \frac{\partial f}{\partial y}. \text{ In particular, show that}$$

$$\frac{\partial}{\partial x} \frac{\partial f}{\partial y} = 1 \neq -1 = \frac{\partial}{\partial y} \frac{\partial f}{\partial x}. \tag{2.16}$$

We thus encounter the fact that the order in which one takes second (or more) derivatives may produce different results. The following result shows that continuity "repairs this situation":

Theorem 2.29   *Suppose $f : \mathcal{D}(f) \subset \mathbb{R}^m \to \mathbb{R}^1$ and there exists an open ball B, centered at x_0, such that $B \subset \mathcal{D}(f)$ and $D_j f, D_k f, D_j D_k f, D_k D_j f$ exist for all points in B and are continuous at x_0. Then $(D_k D_j f)(x_0)$ equals $(D_j D_k f)(x_0)$.*

Observation: We assumed, in Theorem 2.29, that f is real-valued, we did this for simplicity; if f had values in $\mathbb{R}^n$ the result of Theorem 2.29 is still valid and it follows immediately from the more restricted one by applying it to each coordinate of f. This observation applies in many other cases.

Proof: Let $g = f(\cdot, y + h) - f(\cdot, y)$, where $x, y \in B$ and h, k are small numbers. By the Mean Value Theorem, see (2.14′), there exists ξ between x and $x + h$ such that $g(x + h) - g(x) = g'(\xi)h$. Hence,

$$\begin{aligned} & f(x + h, y + k) - f(x + h, y) - f(x, y + k) + f(x, y) \\ &= g(x + h) - g(x) \\ &= [D_1 f(\xi, y + k) - D_1 f(\xi, y)]h \\ &= D_2 D_1 f(\xi, \eta)kh \end{aligned}$$

with η between y and $y + k$.

Now apply the same procedure to $\tilde{g} = f(x + h, \cdot) - f(x, \cdot)$ and we find σ, τ between $x, x + h$ and $y, y + k$ such that

$$(D_2 D_1 f)(\xi, \eta)kh = (D_1 D_2 f)(\sigma, \tau)kh.$$

Cancel kh and let $k, h \to 0, 0$. The continuity assumption gives us the desired result.   ∎

(HW) 2.30   Assume $f : B \to \mathbb{R}^1$, where B is an open ball about $x_0 \in \mathbb{R}^5$ and all partial derivatives of order ≤ 3 exist at each point of B and are continuous in B. How many different values (at most) can these derivatives have at x_0?

We shall show how quite general, sufficiently smooth, functions can be approximated by polynomials.

We begin with the one dimensional case. Suppose f is a real-valued function defined on an open interval $I = (a, b)$ and $x_0 \in I$. Let us assume that f is m times differentiable at x_0; that is, $f'(x_0), f''(x_0), \cdots, f^{(m)}(x_0)$ exist.[2] Consider the polynomial

$$P(x) = \sum_{k=0}^{m} \frac{f^{(k)}(x_0)}{k!}(x - x_0)^k. \tag{2.17}$$

(HW) 2.31   Show that P is the unique polynomial of degree $\leq m$ satisfying $P^{(k)}(x_0) = f^{(k)}(x_0)$ for $k \leq m$.

The polynomial P in (2.17) is called the *Taylor polynomial* (of f) *of degree m at x_0*. It is denoted by $T_{x_0}^m f$. The following result shows that how $T_{x_0}^m f$ approximates f near x_0:

Theorem 2.32 Taylor's Formula.   *Suppose $f, f', \cdots, f^{(m+1)}$ all exist on the open interval $I = (a, b)$ and $x, x_0 \in I$, then there exists a point ξ between x and x_0 such that*

$$f(x) = (T_{x_0}^m f)(x) + \frac{f^{(m+1)}(\xi)}{(m + 1)!}(x - x_0)^{m+1}. \tag{2.18}$$

Comment. Observe that when $m = 0$ this is precisely the Mean Value Theorem (see (2.14')).
Proof: Fix $x \neq x_0$ (both points in I) and let $\mathcal{K}$ satisfy

$$f(x) = \sum_{k=0}^{m} \frac{1}{k!} f^{(k)}(x_0)(x - x_0)^k + \mathcal{K}(x - x_0)^{m+1}.$$

Define $g(t)$ for t in a closed interval in I including x and x_0 by letting

$$g(t) = -f(x) + \sum_{k=0}^{m} \frac{1}{k!} f^{(k)}(t)(x - t)^k + \mathcal{K}(x - t)^{m+1}.$$

---

[2]We are using the common notation $f^{(0)}(x) = f(x)$, $f^{(k)}(x) = \dfrac{d^k f}{dx^k}(x)$, $1 \leq k$, and $f'(x) = \dfrac{df}{dx}$, $f''(x) = \dfrac{d^2 f}{dx^2}$.

Clearly $g(x_0) = 0$ and $g(x) = 0$. Moreover,

$$g'(t)$$

$$= \sum_{k=0}^{m} \frac{1}{k!} f^{(k+1)}(t)(x-t)^k - \sum_{k=1}^{m} \frac{1}{(k-1)!} f^{(k)}(t)(x-t)^{k-1} - (m+1)\mathcal{K}(x-t)^m$$

$$= \frac{1}{m!} f^{(m+1)}(t)(x-t)^m - (m+1)\mathcal{K}(x-t)^m.$$

Again we use the Mean Value Theorem (2.14') (in this case, since $g(x_0) = 0 = g(x)$, this is *Rolle's Theorem*) and conclude that there exists ξ between x and x_0 such that $g'(\xi) = 0$. This allows us to solve for $\mathcal{K}$: $\mathcal{K} = \dfrac{1}{(m+1)!} f^{(m+1)}(\xi)$. ■

Comment: We have seen several applications of the Mean Value Theorem. This is good evidence that it is an important result in the Calculus. This is also not surprising since, as we have already observed, it is easy to see how it is closely related to the derivative and, locally, to a linear approximation to a general function. There are various facts that annoy me about the world nowadays, not only in politics but, also, about how mathematics teaching is developing. We will refrain from saying much about this. Nevertheless, let us observe that there appears to be a tendency, recently, to minimize the role of the Mean Value Theorem in modern texts and courses in the Calculus. We say "Grumph" about this.

The last term in equality (2.18) is referred to as an *error term* in Taylor's formula. The following exercises show how it can be used in applications.

(HW) 2.33   Show that $e = 1 + 1 + \dfrac{1}{2} + \dfrac{1}{6} + \dfrac{1}{2}\dfrac{1}{120} + \cdots = \dfrac{326}{120} + \cdots = 2.71666\cdots$ within 0.005 (convince yourself that $e^\xi < 3$ when $\xi \in (0, 1)$).

(HW) 2.34   Calculate $\sin 36°$ within an error no bigger than 0.001.

Let us now pass to the higher dimensional case of Taylor's formula. This is an easy application of the one dimensional case and the differentiation of composite functions. Let f map an open ball about $x_0 \in \mathbb{R}^n$ into $\mathbb{R}$ and consider the following formal argument (we will state a precise result shortly). Let $g(t) = f(x_0 + t\,h)$ for t in a (small) open interval about $0 \in \mathbb{R}^1$. Then, using Theorem 2.32, we have (for $h \in \mathbb{R}^n$)

$$f(x_0 + h) = g(1) = \sum_{k=0}^{m} \frac{1}{k!} g^{(k)}(0)(1-0)^k + \frac{g^{(m+1)}(\tau)(1-0)^{(m+1)}}{(m+1)!} \qquad (2.19)$$

for a $\tau \in (0,1)$. Let us calculate the derivatives of g using the result obtained in exercise 40. Observe that $g = f \circ \varphi$, where $\varphi(t) = x_0 + t h$. Thus,

$$g'(t) = \sum_{j=1}^{n} D_j f(x_0 + t h) \varphi_j'(t) = \sum_{j=1}^{n} D_j f(x_0 + t h) h_j.$$

$$g''(t) = \sum_{i,j=1}^{n} D_i D_j f(x_0 + t h) h_i h_j$$

and, in general,

$$g^{(k)}(t) = \sum_{i_1,i_2,\cdots,i_k=1}^{n} D_{i_1} D_{i_2} \cdots D_{i_k} f(x_0 + t h) h_{i_1} h_{i_2} \cdots h_{i_k} \qquad \text{for } k \in \mathbb{N}.$$

Different authors use different notations for the expressions of these derivatives. Let us agree to consider the operator $\sum_{j=1}^{n} h_j D_j$ that maps f into

$\sum_{j=1}^{n} h_j D_j f(x_0 + t h)$ and its iterates $(\sum_{j=1}^{n} h_j D_j)^k$ that map f into the last expression of $g^{(k)}(t)$, so that equality (2.19) becomes

$$f(x_0 + h) = \sum_{k=0}^{m} \frac{1}{k!} (\sum_{j=1}^{n} h_j D_j)^k f(x_0) + \frac{1}{(m+1)!} (\sum_{j=1}^{n} h_j D_j)^{(m+1)} f(\xi), \qquad (2.20)$$

where ξ lies on the line segment joining x_0 to $x_0 + h$.

Suppose $h = (h_1, \cdots, h_n) \neq 0$; then the sum $\sum_{j=1}^{n} (h_j D_j f)(x_0 + t h)$ is the directional derivative $(D_h f)(x_0)$ in the direction h (see (2.11)).[1] Also $\sum_{j=1}^{n} h_j D_j = <h, \nabla >$ (formally) and

$$(\sum_{j=1}^{n} h_j D_j f)(x_0) = <h, \nabla > f(x_0). \qquad (2.21)$$

Earlier we stated that we wanted to define the classes $C^m = C^m(B)$ for real-valued functions defined on an open ball $B \subset \mathbb{R}^m$. We do this by saying that f is of class $C^m = C^m(B)$, $m \geq 1$,

---

[1] We defined this notion for unit vectors originally; it clearly applies to any non-zero vector h.

if f and its differential df are of class $C^{m-1}(B)$ ($f \in C^0(B)$ if it is continuous on B). That is, f and all its partial derivatives exist and are continuous (up to order m).

Theorem 2.35 Taylor's Formula in $\mathbb{R}^n$. *If $f : B \subset \mathbb{R}^n \to \mathbb{R}^1$, $B = B_r(x_0)$, $r > 0$, is of class $C^{m+1}(B)$ and $x \in B$, then there exists a point ξ on the line segment joining x to x_0 such that*

$$f(x) = \sum_{k=0}^{m} \frac{1}{k!}(< x - x_0, \nabla >)^k f(x_0) + \frac{1}{(m+1)!}(< x - x_0, \nabla >)^{m+1} f(\xi). \tag{2.22}$$

Because of our assumption that $f \in C^{m+1}(B)$ the calculations that led to equality (2.20) are legitimate and we have, indeed, proved Theorem 2.35. By Theorem 2.29 many of the derivatives resulting from the operators $(< x - x_0, \nabla >)^k$ are equal to each other and, therefore, can be entered in formula (2.22) as a mutiple of a derivative of the form $\dfrac{\partial}{\partial x_{i_1}} \dfrac{\partial}{\partial x_{i_2}} \cdots \dfrac{\partial}{\partial x_{i_k}}$ with $i_1 \leq i_2 \leq \cdots \leq i_k$. For example,

if $n = 2$ and $x_0 = (a, b)$, $x = (a + h, b + k)$, then (2.22) can be written in the form

$$f(a + h, b + k)$$
$$= \sum_{\ell=0}^{m} \frac{1}{\ell!}(h\frac{\partial}{\partial x} + k\frac{\partial}{\partial y})^\ell f(a, b) + \frac{1}{(m+1)!}(h\frac{\partial}{\partial x} + k\frac{\partial}{\partial y})^{m+1} f(\xi, \eta). \tag{2.23}$$

Thus, the term $k = 2$ is

$$\frac{1}{2!}(h^2\frac{\partial^2 f}{\partial x^2}(a, b) + 2hk\frac{\partial^2 f}{\partial x \partial y}(a, b) + k^2\frac{\partial^2 f}{\partial y^2}(a, b)).$$

(HW) 2.36   Use the binomial theorem and Theorem 2.29 to express the kth term in the sum in (2.23) if $f \in C^{m+1}(B)$ when B is an open disk (in $\mathbb{R}^2$) about the point (a, b).

(HW) 2.37   Write out Taylor's fomula for $f(x, y) = \sin x \sin y$ when $x_0 = (0, 0)$.

(HW) 2.38   Let

$$f(x) = \begin{cases} e^{-\frac{1}{x}} & x > 0 \\ 0 & x \leq 0 \end{cases}.$$

Show that $f \in C^m(\mathbb{R}^1)$ for all $m \geq 0$. Write out Taylor's formula for each m when $x_0 = 0$. What is the Taylor's polynomial $(T_0^m f)(x)$? How close is it to $f(x)$ when $x > 0$. Any comments about this?

Taylor's formula shows that a sufficiently smooth (belongs to C^m) function can be approximated by polynomials at least locally. We indicated in (HW) 2.30 and (HW) 2.31 some applications of this fact. In general, however, this is not a very "efficient" approximation and it does require more smoothness for better approximations. If a function is merely continuous on a closed interval (or on a closed rectangle in $\mathbb{R}^m$) it can be approximated by polynomials arbitrarily well.

We present a result by Karl Weierstrass obtained in the middle of the 19th century:

Theorem 2.39 Weierstrass Approximation Theorem.   *Suppose f is a real-valued continuous function on an interval $I \subset \mathbb{R}$. Then, on each bounded, closed subinterval $J \subset I$, there exists a sequence $\{p_n\}$ of polynomials such that $p_n \to f$ uniformly.*

Proof: Let us first suppose f is defined on the entire real line and $f(x) = 0$ if $|x| \geq \frac{1}{2}$. Let

$$q_n(x) = c_n(1 - x^2)^n$$

where c_n is determined by the condition

(∗)
$$\int_{-1}^{1} q_n(y)dy = 1.$$

We then define

$$p_n(x) = \int_{-\infty}^{\infty} f(y)q_n(x - y)dy = \int_{-\infty}^{\infty} f(x - y)q_n(y)dy. \qquad (2.24)$$

Observe that these are not improper integrals since $f \equiv 0$ outside $[-1, 1]$.

We claim that each p_n is a polynomial and $\lim_{n \to \infty} p_n(x) = f(x)$ uniformly on $[-1, 1]$.

From the first equality in (2.24) we see immediately that p_n is a polynomial. If $|x| \leq \frac{1}{2}$, since $f(x - y) = 0$ if $|x - y| > \frac{1}{2}$ it follows that the integral after the second equality in (2.24) is

(∗∗)
$$\int_{-1}^{1} f(x - y)q_n(y)dy = p_n(x)$$

(since $-\frac{1}{2} \leq x \leq \frac{1}{2}$ and $-\frac{1}{2} \leq y - x \leq \frac{1}{2}$ imply $-1 \leq y \leq 1$ by adding the 1st two inequalities). Thus, by (∗) and (∗∗),

(†)
$$f(x) - p_n(x) = f(x) \int_{-1}^{1} q_n(y)dy - \int_{-1}^{1} f(x - y)q_n(y)dy$$

$$= \int_{-1}^{1} [f(x) - f(x - y)]q_n(y)dy.$$

Suppose $\epsilon > 0$ and $M = \max\{|f(x)| : x \in \mathbb{R}\}$ (which is $< \infty$ since $|f(x)| > 0$ only on $[-\frac{1}{2}, \frac{1}{2}]$ and, thus, by Corollary 1.48, $M < \infty$). By the uniform continuity of f (convince yourself of this), we can find $\delta > 0$ such that $|f(x) - f(x - y)| < \epsilon$ if $|y| < \delta$. Using these facts, that $q_n(y) \geq 0$ if $|y| \leq 1$ and (†), we then have

$$|f(x) - p_n(x)| \leq \int_{|y| < \delta} \epsilon q_n(y)dy + \int_{\delta \leq |y| \leq 1} 2M q_n(y)dy$$

$$\leq \epsilon \cdot 1 + 2M \int_{\epsilon \leq |y| \leq 1} q_n(y)dy.$$

The desired result in the special case we are considering will be true if we show that, for each $\delta > 0$,

$$(\star) \qquad \qquad \lim_{n \to \infty} \int_{\delta \leq |y| \leq 1} q_n(y)dy = 0.$$

To see this choose r, $0 < r < 1$. By the definition of c_n,

$$\frac{1}{c_n} = \int_{-1}^{1} (1 - y^2)^n dy > \int_{-r}^{r} (1 - y^2)^n dy \geq \int_{-r}^{r} (1 - r^2)^n dy > 2r(1 - r^2)^n.$$

Hence, $c_n \leq \dfrac{1}{2r(1 - r^2)^n}$ and

$$\int_{\delta \leq |y| \leq 1} q_n(y)dy \leq \frac{1}{2r(1 - r^2)^n} \int_{-1}^{1} (1 - \delta^2)^n dy = \frac{(1 - \delta^2)^n}{r(1 - r^2)^n}.$$

If we also choose $r < \delta$, then the expression on the right tends to 0 as n tends to ∞; thus, $(\star)$ is true.   ∎

The general case is not hard to establish by using the result we just established. We leave the rest of the proof as an exercise for which we give appropriate hints:

(HW) 2.40   Write the proof for the general case for Theorem 2.39. You can use the following suggestions: First assume f vanishes for $x \notin [-r, r]$. Dilate f by forming the function $g(x) \equiv f(2rx)$. Then g vanishes for $x \notin [-\frac{1}{2}, \frac{1}{2}]$ and, by the result just obtained, we know that for each $\epsilon > 0$ we can find a polynomial p_ϵ such that $|p_\epsilon(x) - g(x)| < \epsilon$ if $|x| \leq \frac{1}{2}$. Then, $p_\epsilon(\frac{x}{2r})$ is a polynomial q_ϵ such that $|q_\epsilon(x) - f(x)| < \epsilon$ for $|x| \leq r$. A translation then gives us the result for any continuous f vanishing outside a bounded closed interval I. It is also useful to observe that if f is continuous on the open interval (a, b), $-\infty < a < b < \infty$, we can construct φ_n that is 1 on $[a + \frac{1}{n}, b - \frac{1}{n}]$ and vanishes identically near a and b and φ_n is continuous. We can then find a polynomial p_n (by what we have just shown) $\ni |p_n(x) - \varphi_n(x) f(x)| < \frac{1}{n}$ on (a, b). It follows that $\{p\}$ approximates f uniformly on each closed bounded subinterval of (a, b). Now get to work and write out a proof of Theorem 2.39!

(HW) 2.41   Draw pictures of the polynomials q_n. The silly physicists talk about the *Dirac δ function at x_0*. They say that it is the "function that is 0 at $x \neq x_0$ and has value ∞ at x_0 in such a way that $\int_{-\infty}^{\infty} \delta_{x_0}(x) dx = 1$ (where δ_{x_0} is this function)." Clearly there is no such function. They say even stranger things:

$$\int_{-\infty}^{\infty} \delta_{x_0}(x) f(x) dx = f(x_0) \tag{2.25}$$

if f is continuous on $\mathbb{R}$. See if you can interpret the sequence $\{q_n\}$ as "representing" the Dirac Delta function at $x_0 = 0$.

The Weierstrass approxmation theorem has many extensions. In n-dimensions we can uniformly approximate continuous functions on closed rectangles by polynomials in $x_1, x_2, \cdots, x_n$. We will not present a proof of this and other, more general results (we can't do everything).

2.3   MAXIMA AND MINIMA FOR REAL VALUED FUNCTIONS OF SEVERAL VARIABLES

Let us consider real valued functions f of n real variables. For simplicity we assume the domain of f to be $\mathbb{R}^n$ (though most of what we say will be valid for domains that are open connected sets). Suppose f is differentiable at a so that

$$\lim_{h \to 0, h \neq 0} \frac{f(a + h) - f(a) - <\nabla f(a), h>}{\|h\|} = 0. \tag{2.26}$$

This tells us that $f(a + h)$, as a function of h, is well approximated by the *affine* function $g(h) = f(a) + <\nabla f(a), h>$ (many call this a linear function). We discussed this earlier. Let us look at this geometrically. If $n = 2$, the graph of the equation $z = f(x, y)$ represents a surface in $\mathbb{R}^3$. In Fig. 2.2, the plane represents the graph of $(x, y) \to g(x - x_0, y - y_0) = f(x_0, y_0) + <\nabla f(x_0, y_0), (x - x_0, y - y_0)>$ and the graph of the surface $z = f(x, y)$. Both graphs contain the point $(a, f(a))$ and, at nearby points, $z_{surface} - z_{plane} = f(a + h) - g(h)$ and this distance tends to 0 faster than $\|h\|$ as $h \to (0, 0)$. This represents the meaning of the *tangent plane at a*. With a little stretch of imagination we can consider, in the n-dimensional case, that $z = f(x)$ represents a "hyper-surface" in $\mathbb{R}^{n+1}$ having coordinates $(x_1, \cdots, x_n, z)$ and $z = f(a) + (\nabla f(a), (x - a))$ represents its *tangent hyperplane at a*.

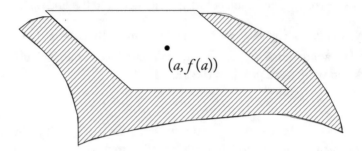

Figure 2.2: Tangent plane to the smooth surface $z = f(x, y)$ at the point $(a, f(a)) = (x_0, y_0, z_0)$.

Let us examine the role played by the gradient at *local maxima* and *local minima* at a (that is, points $a \in \mathcal{D}(f)$ such that $f(x) \leq f(a)$ or $f(x) \geq f(a)$ for x in a neighborhood of a).

Theorem 2.42   *If f has a local maximum at a (or minimum) and f is differentiable at a, then* $\nabla f(a) = 0$.

(HW) 2.43   Prove this theorem (use the fact that if u is any direction, then (see (2.11))

$$(D_u f)(a) = < (\nabla f)(a), u > . \tag{2.27}$$

You undoubtedly know the 1-dimensional version of this theorem. The fact that $f'(a) = 0$, in this case, does not imply that a is a local maximum or minimum. You probably recall that there is a "second derivative" test (and other related results) that can be used to identify these maxima and minima. We shall now explore this situation in higher dimensions.

Definition 2.44   *If f is differentiable in a neighborhood of a. Then a is called a critical point of f iff* $\nabla f(a) = 0$.

(HW) 2.45

(a) Find the critical points of the function $f(x, y) = xy(12 - 3x - 4y)$;

(b) Do the same for $f(x, y) = y^3 - 3x^2 y$.

You should have shown that the origin is a critical point for each of the functions in the last exercise. In each case, however, this point is neither a local minimum nor a maximum. In the first case f is, locally, positive in the open first and third quadrant and negative in the open second and fourth quadrant. Sketch the graph of $z = f(x, y)$ near $(0, 0)$ and you will see that it resembles a saddle (the rider's legs hang down the second and fourth quadrant). For the second function the lines $y = 0$, $y = \sqrt{3}x$ and $y = -\sqrt{3}x$ separate the plane into six regions on which f is alternatively positive and negative. This provides places for a monkey, with a long tail hanging down, to sit on the graph at 0 so that its legs and tail hang down. We thus have a "*monkey saddle* point at 0".

Let us recall the one dimensional situation concerning finding conditions that tell us when a function has a local maximum or minimum, or neither, at a critical point. The simple example $f(x) = x^3$ shows that 0 is a critical point but neither a local maximum or minimum. If f is of class C^2 and a is a critical point, then, if $f''(a) > 0$ $(f''(a) < 0)$ the point a is a local minimum (maximum). Unfortunately, if $f''(a) = 0$ we cannot draw any conclusion. This "second derivative test" has an extension to several variables.

Let us consider the two dimensional case. Suppose f is in the class C^3 in a neighborhood of $a = (r, s) \in \mathbb{R}^2$. We can then write the Taylor formula (2.23) in the form

$$
\begin{aligned}
f(r + h, s + k) &= f(r, s) + h D_1 f(r, s) + k D_2 f(r, s) + \frac{1}{2}[h^2 D_1 D_2 f(r, s) \\
&\quad + hk D_1 D_2 f(r, s) + kh D_2 D_1 f(r, s) + k^2 D_2 D_2 f(r, s)] \\
&\quad + \frac{1}{3!}(h D_1 + k D_2)^3 f(r + \theta h, s + \theta k) \\
&= f(r, s) + < \triangle f(r, s), (h, k) > \\
&\quad + \frac{1}{2} < (h, k), \begin{pmatrix} D_1 D_1 f(a) & D_1 D_2 f(a) \\ D_1 D_2 f(a) & D_2 D_2 f(a) \end{pmatrix} \begin{pmatrix} h \\ k \end{pmatrix} > + E
\end{aligned}
$$

where we used the fact that $D_1 D_2 f(a) = D_2 D_1 f(a)$ (by Theorem 2.29) and E is the (third order) error term. The matrix

$$
H = H_f(a) = \begin{pmatrix} D_1 D_1 f(a) & D_1 D_2 f(a) \\ D_2 D_1 f(a) & D_2 D_2 f(a) \end{pmatrix}
$$

is called the *Hessian of f at a*. In our case, as we just pointed out, it is symmetric. That is, it has the form

$$
\left[\begin{pmatrix} D_1 D_1 & D_1 D_2 \\ D_2 D_1 & D_2 D_2 \end{pmatrix} f \right] (a) = \begin{pmatrix} \alpha & \beta \\ \beta & \gamma \end{pmatrix} = H = H_f(a). \tag{2.28}
$$

There is a theorem in linear algebra that asserts that, for each such symmetric matrix, one can find an orthonormal basis, with respect to which, the linear transformation, produced by this matrix, has a diagonal matrix $\begin{pmatrix} \lambda_1 & 0 \\ 0 & \lambda_2 \end{pmatrix}$. If we use this result, it is fairly easy to show the following result:

Theorem 2.46 *Suppose $f \in C^2$ in an open subset in $\mathbb{R}^2$ that contains a point a such that $\nabla f(a) = 0$. Let $\alpha = f_{11}(a)$, $\beta = f_{12}(a) = f_{21}(a)$, $\gamma = f_{22}(a)$. Then*

(a) *If $\alpha\gamma - \beta^2 < 0$, f has a saddle point at a;*

(b) *If $\alpha\gamma - \beta^2 > 0$, and $\alpha > 0$, f has a local minimum at a;*

(c) *If $\alpha\gamma - \beta^2 > 0$, and $\alpha < 0$, f has a local maximum at a;*

(d) *If $\alpha\gamma - \beta^2 = 0$, no conclusion can be drawn.*

We shall not prove this theorem, nor will we state an extension to higher dimensions. In most particular cases (as those in (HW) 2.41) one can figure out the behavior of an f at a critical point.

We finish this section with an observation about the gradient of a smooth function (say, class C^1) at any point. Let u be a direction (a unit vector) in $\mathbb{R}^n$ and $x_0 \in \mathcal{D}(f)$; then the directional derivative $(D_u f)(x_0)$ equals

$$(D_u f)(x_0) = <\nabla f(x_0), u>$$

(see (2.27)). Schwarz's inequality tells us that $| < \nabla f(x_0), u > | \le \|(\nabla f)(x_0)\| \|u\| = \|\nabla f(x_0)\|$. Suppose $\nabla f(x_0)$ is not zero and u is *the direction of* $\nabla f(x_0)$, that is,

$$u = \frac{\nabla f(x_0)}{\|\nabla f(x_0)\|}.$$ We then have equality in Schwarz's inequality:

$$| < \nabla f(x_0), u > | = \|\nabla f(x_0)\|. \tag{2.29}$$

Thus, $\|\nabla f(x_0)\|$ is the largest absolute value of all the directional derivatives of f at x_0. Think of the graph of f as representing a region of mountains and valleys (best to do this in $\mathbb{R}^3$). If you are standing at a point x_0 the vector $\nabla f(x_0)$ points in the direction of steepest decent or ascent. If you desire to walk without going up or down, just go in the direction where the directional derivative is 0, which is orthogonal to $\nabla f(x_0)$. The "level lines" you see on maps are curves along which these directional derivatives are zero.

(HW) 2.47   Find the absolute maximum and minimum values

$$f(x, y) = \frac{x}{x^2 + (y-1)^2 + 4}$$

on the first quadrant $S = \{(x, y) : x, y \ge 0\}$. Be careful, S is not an open set.

2.4   THE IMPLICIT FUNCTION THEOREM

Let $F(x, y) = F(x_1, x_2, \cdots, x_n, y) = \alpha_1 x_1 + \cdots + \alpha_n x_n + \beta y + c$. If $a = (\alpha_1, \cdots, \alpha_n)$ the equation $F(x, y) = 0$ has the solution $y = -\dfrac{<x, a>}{\beta} - c = f(x)$ iff $\beta \neq 0$. We have seen that any C^1 function $F(x, y)$ is approximately linear. More precisely, $F(x, y) = F(a, b) + <\nabla F(a, b), x - a> + \dfrac{\partial}{\partial y} F(a, b)(y - b) +$ small error for $(x, y) = (x_1, \cdots, x_n, y)$ near $(a, b) = (a_1, \cdots, a_n, b) \in \mathbb{R}^{n+1}$. Thus, if $F(a, b) = 0$ it is not unreasonable to expect that the equation $F(x, y) = 0$ can be solved near (a, b) provided $\dfrac{\partial F}{\partial y}(a, b) \neq 0$. That is, there exists a function $f(x)$ such that $F(x, f(x)) = 0$ for x in a neighborhood of a. In fact, we have

Theorem 2.48   *(The Implicit Function Theorem for a Single Equation). Let $F(x, y) = F(x_1, \cdots, x_n, y)$ be a C^1 function on a neighborhood of a point $(a, b) = (a_1, \cdots, a_n, , b) \in \mathbb{R}^{n+1}$ for which we have $F(a, b) = 0$ and $\dfrac{\partial F}{\partial y}(a, b) \neq 0$. Then there exist $r_0 > 0$ and $r_1 > 0$ such that*

(i) for each $x \in B_{r_0}(a) = \{w \in \mathbb{R}^n : \|w - a\| < r_0\}$ there exists a unique y such that $|y - b| < r_1$ for which $F(x, y) = 0$. Let $y = f(x)$ (in particular, $f(a) = b$);

(ii) the function f is of class C^1 in $B_{r_0}(a)$ and its partial derivatives satisfy

$$\frac{\partial f}{\partial x_j}(x) = -\frac{\dfrac{\partial F}{\partial x_i}(x, f(x))}{\dfrac{\partial F}{\partial y}(x, f(x))}, \tag{2.30}$$

for $j = 1, 2, \cdots, n$.

Remark: Formula (2.30) follows by the *"chain rule"* (Theorem 2.25) obtained by differentiating the equation $F(x, f(x)) = 0$.

Before going into the proof of this theorem let us examine some examples.

(HW) 2.49   Let $G(x, y) = x - e^{1-x} - y^2$. Say all you can about solutions f and g such that $G(x, f(x)) = 0$ and $G(g(y), y) = 0$.

The following example is more challenging:

The *lemniscate* (see Fig. 2.3) is the locus of all points $(x, y) \in \mathbb{R}^2$ for which the product of the distances r_1 and r_2 from the points $(-a, 0)$ and $(a, 0)$ has the value a^2. Since $r_1^2 = (x - a)^2 + y^2$ and $r_2^2 = (x + a)^2 + y^2$, the equation of the lemniscate is

$$(x^2 + y^2)^2 - 2a^2(x^2 - y^2) = F(x, y) = 0. \tag{2.31}$$

$\dfrac{\partial F}{\partial y}(x, y) = 4y(x^2 + y^2 + a^2)$. Hence, $\dfrac{\partial F}{\partial y}(x_0, y_0) \neq 0$ iff $y_0 \neq 0$. According to Theorem 2.48 we know that there exists a unique f on an open interval about x_0 such that $F(x, f(x)) = 0$ (x in this interval) and

$$f'(x) = -\frac{\dfrac{\partial F}{\partial x}(x, f(x))}{\dfrac{\partial F}{\partial y}(x, f(x))} = -\frac{x}{y}\frac{x^2 + y^2 - a^2}{x^2 + y^2 + a^2}.$$

This formula gives us a lot of information: The critical points are at $x = \pm\dfrac{a}{2}\sqrt{3}$, the maximal and minimal value of f (at these points) is $\dfrac{a}{2}$ and $-\dfrac{a}{2}$, x's range is $(-\sqrt{2}a, \sqrt{2}a)$, $\cdots$.

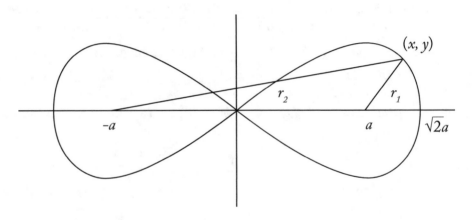

Figure 2.3: The lemniscate.

This last case shows that we can say a lot about f even when we do not have an explicit expression for this function.

Proof of Theorem (2.48): We first establish (i). We can assume $\dfrac{\partial F}{\partial y}(a, b) > 0$. By continuity $\dfrac{\partial F}{\partial y}$ remains positive in a neighborhood of (a, b): say, $\{(x, y) : \|x - a\| < r_1, |y - b| < r_1\}$. Hence, on this set, $F(x, y)$ is strictly increasing, as a function of y, for each x. Since $F(a, b) = 0$, $F(a, b + r_1) > 0$ and $F(a, b - r_1) < 0$. The continuity if F implies that $F(x, b + r_1) > 0$ and $F(x, b - r_1) < 0$ for $\|x - a\| < r_0$ for some positive $r_0 \leq r_1$. It follows from the intermediate value property (state it!) that there exists a unique y for each $x \in B_{r_0}(a)$ satisfying $|y - b| < r_1$ and $F(x, y) = 0$. This establishes (i).

The function f such that $f(x) = y$, we just defined, must be continuous at $x = a$. We have, in fact, shown that $|f(x) - f(a)| = |y - b| < r_1$ if $\|x - a\| < r_0$. If we repeat the same argument with r_1 replaced by ϵ (should $\epsilon < r_1$) we then obtain a δ, instead of r_0 such that $\|x - a\| < \delta$. In fact, it is easy to show continuity of f at any other $x_0 \in B_{r_0}(a)$.

We next show that each $\dfrac{\partial f}{\partial x_i}$ exists in $B_{r_0}(a)$ and are given by (2.30) (hence, they are continuous). If $x \in B_{r_0}(a)$ and $h = (0, \cdots, 0, \alpha, 0, \cdots, 0)$ (α small), so that $x + h \in B_{r_0}(a)$, let $y = f(x), k = f(x + h) - y$. Then $F(x + h, y + k) = F(x, y) = 0$ and, by the mean value theorem,

$$0 = F(x + h, y + k) - F(x, y) = h\frac{\partial F}{\partial x_j}(x + th, y + tk) + k\frac{\partial F}{\partial y}(x + th, y + tk)$$

for some $t \in (0, 1)$. From this equality we obtain

$$\frac{f(x + h) - f(x)}{\alpha} = \frac{k}{\alpha} = -\frac{\dfrac{\partial F}{\partial x_j}(x + th, y + tk)}{\dfrac{\partial F}{\partial y}(x + th, y + tk)}.$$

Now let $\alpha \to 0$ ($\therefore h \to 0$). Since f is continuous $k \to 0$. Since $\dfrac{\partial F}{\partial x_j}, \dfrac{\partial F}{\partial y}$ are continuous and $\dfrac{\partial F}{\partial y} \neq 0$ the passage to the limit gives us (2.30). ∎

Corollary 2.50   *(The Inverse Function Theorem) Let $g : I \subset \mathbb{R} \to \mathbb{R}$ be a differentiable function on the (non-empty) open interval I and assume $g'(y) \neq 0$ for all $y \in I$. Then the image, J, of I is an interval on which a unique function f is defined such that $g(f(x)) = x$. The function f is differentiable on J and*

$$f'(x) = \frac{1}{g'(f(x))}.$$

(HW) 2.51   Write out a proof of Corollary 2.50 by applying Theorem 2.48 to the function $F(x, y) = x - g(y)$. Convince yourself (and me) that $f = g^{-1}$ (see page 24 where "inverse function" is defined).

The implicit function theorem for a single equation (Theorem 2.48) extends to $\mathbb{R}^k$-valued functions $F(x, y)$ with $(x, y) = (x_1, \cdots, x_n, y_1, \cdots, y_k) \in \mathbb{R}^{n+k} = \mathbb{R}^n \times \mathbb{R}^k$. It provides conditions that allow us to solve k equations

$$\begin{aligned}
F_1(x_1, \cdots, x_n, y_1, \cdots, y_k) &= 0 \\
&\cdots\cdots\cdots\cdots\cdots\cdots\cdots\cdots \\
F_k(x_1, \cdots, x_n, y_1, \cdots, y_k) &= 0
\end{aligned} \qquad (2.32)$$

for the y's in terms of the x's.

The basic ideas are very similar to the ones we have been using. It is helpful to use vector notation. Thus, let $F(x, y)$ be an $\mathbb{R}^k$-valued C^1 function in a neighborhood of $(a, b) \in \mathbb{R}^{n+k}$. We assume $F(a, b) = 0$ and $det B \neq 0$, where $B = (B_{ij})$ is the $k \times k$ matrix $B_{ij} = \dfrac{\partial F_i}{\partial y_i}(a, b)$. Then we can find $r_0, r_1 > 0$ such that:

(i) For each $x \in B_{r_0}(a)$ there exists a unique $y \in B_{r_1}(b)$ such that $F(x, y) = 0$. We denote this y by $f(x)$ (in particular) $f(a) = b$.

(ii) The function f just defined in $B_{r_0}(a)$ is of class C^1 and $\dfrac{\partial f}{\partial x_j}$ can be computed by differentiating (using the chain rule) $F(x, f(x)) = 0$ with respect to x_j and solving the resulting linear system of equations for $\dfrac{\partial f_1}{\partial x_j}, \dfrac{\partial f_2}{\partial x_j}, \cdots, \dfrac{\partial f_k}{\partial x_j}$.

We will not provide a proof of all this. If time permits this will be discussed in class.

CHAPTER 3

One and Several Dimensional Integral Calculus

3.1 BRIEF REVIEW OF INTEGRALS OF REAL-VALUED FUNCTIONS DEFINED ON A FINITE CLOSED INTERVAL IN $\mathbb{R}$

Suppose $f : [a, b] \to \mathbb{R}$, $-\infty < a < b < \infty$, and f is a bounded non-negative function. You undoubtedly know that the integral $\int_a^b f$ "represents" the area of the region in $\mathbb{R}^2$ that lies under the graph of f and above the interval $[a, b]$ (see Fig. 3.1).

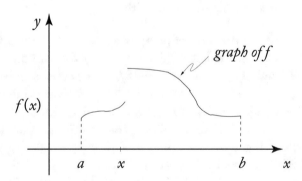

Figure 3.1: The graph of f.

What does this mean? Let us agree that we know what is meant by the area of a rectangle in $\mathbb{R}^2$. Let us partition $[a, b]$ into a finite number of subintervals as follows: $a = x_0 < x_1 < x_2 < \cdots < x_n = b$ so that $I_m = [x_m, x_{m+1}]$, $m = 0, 1, \cdots, n - 1$, cover $[a, b]$ and I_k meets I_{k+1} only at the end point $k + 1$. Since f is bounded ($0 \le f(x) \le M < \infty$ for some positive $M \in \mathbb{R}$) both $u_k = \sup_{x \in I_k} f(x)$ and $\ell_k = \inf_{x \in I_k} f(x)$ exist (and $0 \le \ell_k \le u_k \le M$). Whatever one means by the area below the graph of f and above 0, it is clear that this number A should satisfy

$$\sum_{k=0}^{n-1} \ell_k (x_{k+1} - x_k) \le A \le \sum_{k=0}^{n-1} u_k (x_{k+1} - x_k) \tag{3.1}$$

(see Fig. 3.2). Moreover, we also must have $0 \le A \le M(b-a)$.

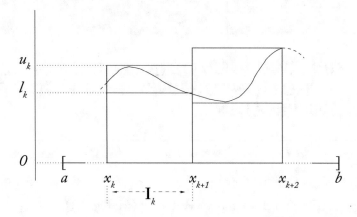

Figure 3.2: Lower and upper bounds of the integral.

Let P be the partition of $[a, b]$ made up of the intervals $I_0, I_1, \cdots, I_{n-1}$. Let us denote by $\ell(P)$ the sum on the left of (3.1) and $u(P)$ the sum on the right; we call $\ell(P)$ the *lower sum* (over P) and $u(P)$ the *upper sum*. If Q is any other such partition, made up by the intervals $J_0, J_1, \cdots, J_{m-1}$, we say that Q is a *refinement* of P if each of the intervals I_j (of the partition P) is a union of some of the intervals J_k (of Q). Any two partitions P and Q have a *common refinement*, $R = P \bigcup Q$, that is obtained as follows: consider the end points $a = x_0 < x_1 < \cdots < x_n = b$ of the intervals of P and the end points $a = y_0 < y_1 < \cdots < y_m = b$ of the intervals of Q. Order the union of the x_j's and y_k's in ascending order and obtain the corresponding ordered sequence $a = z_0 < z_1 < \cdots < z_s = b$ (if $x_j = y_k$ include only one of these points among the z's). Then the partition R made up of the intervals $[z_0, z_1], [z_1, z_2], \cdots, [z_{s-1}, z_s]$ is the common refinement of P and Q. It is clear that

If P and Q are partitions of $[a, b]$ then $\ell(P) \le u(Q)$. \tag{3.2}

Moreover, if Q is a refinement of P, then

$$\ell(P) \le \ell(Q) \le u(Q) \le u(P). \tag{3.3}$$

(HW) 3.1   Prove (3.2) and (3.3).

If P is a partition of $[a, b]$, then the *norm* of P, $\|P\|$, is the LUB of the lengths of the intervals $I_0, I_1, \cdots, I_{n-1}$ making up P. That is,

$$\|P\| = LUB\{x_{j+1} - x_j : j = 0, 1, \cdots, n-1\}.$$

(HW) 3.2   Show that if Q is a refinement of P, then $\|Q\| \leq \|P\|$.

From now on, let us assume that f is bounded, $m \leq f(x) \leq M$ for all $x \in [a, b]$, but not necessarily non-negative. Inequalities (3.2) and (3.3) are still valid in this case. The sets $L = \{\ell(P), P \text{ a partition of } [a, b]\}$ and $U = \{u(P), P \text{ a partition of } [a, b]\}$ are each bounded sets: L is bounded from below by $m(b - a)$ and from above by each $u(Q)$, Q a partition of $[a, b]$ and U is bounded above by $M(b - a)$ and from below by each $\ell(P)$, P a partition of $[a, b]$. It follows that the *sup* = LUB of

$$L \equiv \underline{\int_a^b} f \leq \overline{\int_a^b} f \equiv \inf = \text{GLB of } U. \text{ The numbers}$$

$$\underline{\int_a^b} f \text{ and}$$

$$\overline{\int_a^b} f \text{ are called } \textit{the lower (Riemann) integral} \text{ of } f \text{ on } [a, b] \text{ and } \textit{the upper (Riemann) integral}$$

of f over $[a, b]$.

Definition 3.3   *Suppose* $-\infty < m \leq f(x) \leq M < \infty$ *for all* $x \in [a, b]$. *We say that* f *is (Riemann) integrable over* $[a, b]$ *iff* $\underline{\int_a^b} f = \overline{\int_a^b} f$. *When this is the case, we call this number the integral of* f *and we denote it by* $\int_a^b f$.

Not all bounded functions on a closed finite interval $[a, b]$ are integrable. For example $\underline{\int_0^1} f = 0$ and $\overline{\int_0^1} f = 1$ when

$$f(x) = \begin{cases} 1 \text{ if } x \in [0, 1] \text{ is rational} \\ 0 \text{ if } x \in [0, 1] \text{ is irrational.} \end{cases}$$

A natural question is: Is there a "reasonable" class of functions that are integrable? A rather satisfactory asnswer is the following one:

Theorem 3.4   *Suppose* $[a, b](\subset \mathbb{R})$ *is a finite interval and* f *is a real valued continuous function on* $[a, b]$, *then* f *is integrable on* $[a, b]$. *Remark*: We know (see Corollary 1.48 on page 23) that a continuous function on the compact set $[a, b]$ is bounded. Clearly, an unbounded function cannot be integrable on $[a, b]$.

(HW) 3.5   Show that this last assertion is true.

Proof of Theorem 3.4: We shall show that if $\epsilon > 0$ then

$$0 \le \int_a^{\bar{b}} f - \int_{\underline{a}}^b f < \epsilon. \tag{3.4}$$

The inequality on the left is obvious. In order to obtain the inequality on the right we use Theorem 1.44 that assures us that f is uniformly continuous. Thus, if $\alpha = \epsilon/(b - a) > 0$ there exists $\delta > 0$ such that $|f(x) - f(y)| < \alpha$ if $|x - y| < \delta$. Let P be a partition of norm $\|P\| < \delta$. Then (3.4) follows from

$$\int_a^{\bar{b}} f - \int_{\underline{a}}^b f \le u(P) - \ell(P) = \sum_{k=0}^{n-1} (u_k - \ell_k)(x_{k+1} - x_k) < \alpha(b - a) = \epsilon.$$

The last inequality is true since u_k and ℓ_k are assumed by values $f(x)$ and $f(y)$ for some pair $x, y \in [x_k, x_{k+1}]$ (see Corollary 1.48). Since $|x - y| \le x_{k+1} - x_k \le \|P\| < \delta$, we then have $u_k - \ell_k = f(x) - f(y) < \alpha = \epsilon/(b - a)$. Thus,

$$\sum_{k=0}^{n-1} (u_k - \ell_k)(x_{k+1} - x_k) < \frac{\epsilon}{b - a} \sum_{k=0}^{n-1} (x_{k+1} - x_k) = \frac{\epsilon}{b - a}(b - a) = \epsilon.$$

∎

Given f real valued let f_- and f_+ be the functions

$$f_-(x) = \begin{cases} -f(x) & \text{if } f(x) < 0 \\ 0 & \text{if } f(x) \ge 0 \end{cases}$$

and $f_+ = f + f_-$. Thus, $f(x) = f_+(x) - f_-(x)$ and $|f(x)| = f_+(x) + f_-(x)$.

Continuity is a sufficient condition on $[a, b]$ for integrability of a function on $[a, b]$. It is not, however, a necessary condition. The following result provides us with a class of integrable functions that may have many discontinuities:

Theorem 3.6   *Suppose f is bounded and monotone on $[a, b]$, then f is integrable on this interval.*

Recall that "monotone" means that f is either non-decreasing - - $x \le y \Rightarrow f(x) \le f(y)$ - - or non-increasing - - $x \le y \Rightarrow f(x) \ge f(y)$.

Proof: Suppose f is increasing (or non-decreasing) on $[a, b]$ (a similar argument applies if f is decreasing). We have

$$\int_a^{\bar{b}} f - \int_{\underline{a}}^b f \le u(P) - \ell(P) \tag{3.5}$$

for any partition P of $[a, b]$. Let $P = P_k$ be the partition of $[a, b]$ made up of k subintervals each of length $\dfrac{b-a}{k}$

$$([a, a + \frac{b-a}{k}], [a + \frac{b-a}{k}, a + \frac{2(b-a)}{k}], \cdots, [a + \frac{k-1}{k}(b-a), a + \frac{k}{k}(b-a)]).$$

Then $\ell(P) = \dfrac{b-a}{k} \displaystyle\sum_{j=0}^{k-1} f(x_j)$ and $u(P) = \dfrac{b-a}{k} \displaystyle\sum_{j=1}^{k} f(x_j)$. Thus,

$$u(P) - \ell(P) = \frac{b-a}{k}[f(x_k) - f(x_0)] = \frac{(b-a)[f(b) - f(a)]}{k}.$$

Thus, (3.5) for this partion P is

$$\overline{\int_a^b} f - \underline{\int_a^b} f \le \frac{(b-a)[f(b) - f(a)]}{k}.$$

Letting k tend to ∞, we thus obtain

$$0 \le \overline{\int_a^b} f - \underline{\int_a^b} f \le 0$$

and the desired integrablity is established. ∎

(HW) 3.7 Construct an increasing, bounded f on $[0, 1]$ that has a jump discontinuity at each $\dfrac{1}{n} \in [0, 1]$, $n = 1, 2, 3, \cdots$.

(HW) 3.8 If f is bounded on $[a, b]$ and continuous at all but finitely many points in $[a, b]$, show that f is integrable on $[a, b]$.

Theorem 3.6 and these two exercises show, indeed, that continuity is not a necessary condition for integrability. We shall discuss in class the fact that continuity, however, is "close" to being necessary and sufficient for the integrability of a function on a finite interval $[a, b]$.

We have introduced integrability of a function f that is bounded on a finite interval $[a, b]$ in terms of the upper and lower sums $u(P)$ and $\ell(P)$, P a partition of $[a, b]$. More generally, if P corresponds to $a = x_0 < x_1 < \cdots < x_{n-1} < x_n = b$ and $t_j \in [x_{j-1}, x_j]$, $1 \le j \le n$, we can form the *Riemann sum* of f associated with P:

$$\sum_{j=1}^{n} f(t_j)(x_j - x_{j-1}).$$

It is clear that

$$\ell(P) \le \sum_{j=1}^{n} f(t_j)(x_j - x_{j-1}) \le u(P). \tag{3.6}$$

Thus, if f is integrable and P is a partition for which $\ell(P)$ and $u(P)$ are close to $\int_a^b f$, then any Riemann sum corresponding to P will also be close to $\int_a^b f$.

(HW) 3.9   Prove the following result:

Theorem 3.10   *Suppose* $-\infty < a < b < c < \infty$. *Then,*
(a) if f is integrable on $[a, b]$ and on $[b, c]$ then it is integrable on $[a, c]$ and

$$\int_a^c f(x)dx = \int_a^b f(x)dx + \int_b^c f(x)dx. \tag{3.7}$$

(b) If f and g are integrable on $[a, b]$, then the same is true for $f + g$ and

$$\int_a^b [f + g] = \int_a^b f + \int_a^b g. \tag{3.8}$$

(HW) 3.11   Suppose f and g are integrable on $[a, b]$. Prove the following results:

(i) If $c \in \mathbb{R}$, then cf is integrable on $[a, b]$ and

$$\int_a^b cf = c \int_a^b f; \tag{3.9}$$

(ii)
$$\text{If} \quad [c, d] \subset [a, b] \quad \text{then } f \text{ is integrable on } [c, d]; \tag{3.10}$$

(iii) If g is integrable on $[a, b]$ and $f(x) \le g(x)$ for all $x \in [a, b]$, then

$$\int_a^b f \le \int_a^b g; \tag{3.11}$$

(iv)
$$|f| \text{ is integrable and } \left| \int_a^b f \right| \le \int_a^b |f|. \tag{3.12}$$

The notation $\int_a^b f(x)dx$ is often used for $\int_a^b f$.

The *Fundamental Theorem of Calculus:*

Theorem 3.12   *Let f be integrable on $[a, b]$, $-\infty < a < b < \infty$. By (3.10), f is integrable on $[a, x]$ for each $x \in [a, b]$ and this permits us to consider the function F on $[a, b]$ defined by $F(x) = \int_a^x f$, $x \in [a, b]$. Then F is continuous and $F'(x)$ exists and equals $f(x)$ at each x that is a point of continuity of f.*

Now suppose F is a continuous function on $[a, b]$. Assume, further, that F is differentiable except, possibly, at finitely many points in $[a, b]$. Let f be a function on $[a, b]$ such that $f(x) = F'(x)$ whenever $F'(x)$ is defined. If f is integrable on $[a, b]$, then $\int_a^b f = F(b) - F(a)$.

Proof: If $x, y \in [a, b]$, then, by (3.7), we have

$$F(y) - F(x) = \int_x^y f$$

(we use the convention: $\int_x^y f = -\int_y^x f$; thus, we can have either $x < y$ or $y < x$). Let $C = \sup\{|f(t)| : t \in [a, b]\}$. Using (3.12) we have (assuming $x < y$)

$$|F(y) - F(x)| \le \int_x^y |f| \le C \int_x^y 1 = C|y - x|;$$

that is, $|F(y) - F(x)| \le C|y - x|$ (valid also if $y < x$). This implies uniform continuity of F.

If f is continuous at x, given $\epsilon > 0$, there exists $\delta > 0$ such that $|f(t) - f(x)| < \epsilon$ if $|t - x| < \delta$. We also have

$$f(x) = f(x)\frac{1}{y - x} \int_x^y 1 dt = \frac{1}{y - x} \int_x^y f(x)dt.$$

Consequently,

$$\frac{F(y) - F(x)}{y - x} - f(x) = \frac{1}{y - x} \int_x^y \{f(t) - f(x)\}dt$$

Hence, if $|t - x| < \delta$ we have $|f(t) - f(x)| < \epsilon$ for all t between y and x. It follows that

$$|\frac{F(y) - F(x)}{y - x} - f(x)| \le \frac{1}{|y - x|}|\int_x^y \epsilon dt| = \epsilon.$$

This shows that

$$\lim_{y \to x, y \ne x} \frac{F(y) - F(x)}{y - x} = f(x)$$

as we desired.

We now pass to the proof of the second part Theorem 3.12. Let P be a partition of $[a, b]$ defined by $a = x_0 < x_1 < \cdots < x_n = b$ with $\ell(P)$ and $u(P)$ within $\epsilon > 0$ of $\int_a^b f$. By adding extra points, if needed, we can assume that the points where F is not differentiable are among the x_j's. Then, by the Mean Value Theorem, we can find $t_j \in (x_{j-1}, x_j)$, $j = 1, \cdots, n$, such that $F(x_j) - F(x_{j-1}) = F'(t_j)(x_j - x_{j-1}) = f(t_j)(x_j - x_{j-1})$ (since F is continuous on $[x_{j-1}, x_j]$ and differentiable in (x_{j-1}, x_j)). Adding these equalities we obtain

$$F(b) - F(a) = F(x_n) - F(x_0) = \sum_{j=1}^n f(t_j)(x_j - x_{j-1});$$

moreover,

$$\ell(P) \leq F(b) - F(a) \leq u(P).$$

It follows that $F(b) - F(a)$ is within ϵ of $\int_a^b f$ and the equality $\int_a^b f = F(b) - F(a)$ is true since $\epsilon > 0$ can be made arbitrarily small. ∎

Suppose $f : I = [a, b] \to \mathbb{R}$ and P is the partition associated with $a = x_0 < x_1 < \cdots < x_n = b$. Let

$$v_f(P) = \sum_{i=1}^n |f(x_i) - f(x_{i-1})|.$$

If there exists $M < \infty$ such that $v_f(P) \leq M$ for all partitions P, then f is said to be of *Bounded Variation* on I (BV on I).

(HW) 3.13   Show that

$$f(x) = \begin{cases} x \sin \dfrac{1}{x}, & x \neq 0 \\ 0, & x = 0 \end{cases}$$

is continuous but not of bounded variation.

Theorem 3.14   *If f is of bounded variation on $[a, b]$, then $f = g - h$ where g and h are monotonically nondecreasing functions on $[a, b]$. Conversely, any such f is of BV.*

Proof: The last statement is obvious.

Suppose f is of BV. For any x satisfying $-\infty < a \leq x \leq b < \infty$ let the *total variation* at x be $T_f(x) = T(x)$ be the supremum over all partitions $a = y_0 < y_1 < \cdots < y_n = x$ of the sums $\sum_{i=1}^n |f(y_i) - f(y_{i-1})|$. T is clearly a non-decreasing function and $T(x) \leq T(b)$, the total variation of f on $[a, b]$ (total variation of f at b) which is the *l.u.b.* of all $v_f(P)$ as P ranges through all partitions of $[a, b]$. Consider $h = T - f$ on $[a, b]$. We have, if $a \leq x < y \leq b$,

$$h(y) - h(x) = [T(y) - T(x)] - (f(y) - f(x)).$$

But $T(y) - T(x)$ is the total variation of f on $[x, y]$ and $f(y) - f(x)$ is clearly $\leq T(y) - T(x)$. Thus, h is also non-decreasing and $f = T - h$ is the difference we are looking for with $T = g$. ∎

It follows from Theorems (3.14) and (3.6) that any function of bounded variation on a finite interval $[a, b]$ is integrable on $[a, b]$.

Let us establish some notation and make some comments about the Fundamental Theorem. We have introduced the symbol $\int_a^b f$ to denote the integral of f and also the notation $\int_a^b f(t)dt = \int_a^b f(s)ds = \int_a^b f$. We shall say, when f is defined on $I = (a, b)$, that G is a *primitive of f* if it is differentiable on I and $G'(x) = f(x)$ for all $x \in I$. The fundamental theorem, in particular, tells us that if f is continuous on I, then $F(x) = \int_c^x f$, $c, x \in I$, is a primitive of f; moreover, if G is any other primitive $G(x) - G(c) = F(x)$. We see, therefore, that any two primitives of f differ by a constant.

(HW) 3.15   Suppose f is a continuous function on $I \subset \mathbb{R}$. For each $c \in I$, $F_c(x) = \int_c^x f(t)dt$ is a primitive of f. Give an example for which not all primitives of F are of this form.

(HW) 3.16   If $F(x) = \int_x^a f$ (f continuous on an open interval containing a and x), what is $F'(x)$? If $F(x) = \int_x^{x^2} f$, what is $F'(x)$?

If F is a primitive of f and $G(t) = F(\varphi(t))$ (assume what we need for forming this composition), then the chain rule gives us

$$G'(t) = F'(\varphi(t))\varphi'(t) = f(\varphi(t))\varphi'(t). \tag{3.13}$$

Thus, if $a = \varphi(\alpha)$ and $b = \varphi(\beta)$, then

$$\int_a^b f(x)dx = F(b) - F(a) = G(\beta) - G(\alpha) = \int_\alpha^\beta f(\varphi(t))\varphi'(t)dt. \tag{3.14}$$

Equality (3.14) is known as the *change of variable formula*. You should justify it, based on the discussion preceding it. One usually uses the following language: The change of variable (3.14) is described by the statements

(i) replace x by $\varphi(t)$ and dx by $\varphi'(t)dt$.

(ii) replace the limits a and b by α and β satisfying $a = \varphi(\alpha)$ and $b = \varphi(\beta)$.

This change of variables explains the reason for writing $\int_a^b f(x)dx$ (or $\int_a^b f(u)du$, or $\int_a^b f(\odot)d\odot$). Here is an example of its use:

Calculate $\int_0^1 x\sqrt{1-x^2}dx$. We do this by the change of variable $x = (1-t)^{\frac{1}{2}}$ so that $dx = -\frac{1}{2}(1-t)^{-\frac{1}{2}}dt$. Then,

$$\int_0^1 x\sqrt{1-x^2}dx = -\frac{1}{2}\int_1^0 (1-t)^{\frac{1}{2}}t^{\frac{1}{2}}(1-t)^{-\frac{1}{2}}dt = \frac{1}{2}\int_0^1 t^{\frac{1}{2}}dt.$$

But $\frac{1}{3}t^{\frac{3}{2}}$ is a primitive (sometimes called an *antiderivative*) of $\frac{1}{2}t^{\frac{1}{2}}$. Thus, the last integral equals $\frac{1}{3}t^{\frac{3}{2}}\big|_0^1 = \frac{1}{3}$.

(HW) 3.17   Let $\int f$ denote a (general) primitive of f. Calculate

$$\int \sin^4 x \cos^2 x\, dx.$$

3.2   CURVES, ARC LENGTH, AND LINE INTEGRALS

A function $g : [a,b] \to \mathbb{R}^n$ is called a *parametrization* of a curve, C, in $\mathbb{R}^n$ (n.b. we have not defined a curve, C, in $\mathbb{R}^n$—that will be done shortly). We assume that $[a,b]$ is a finite interval, $g(t) = x(t) = (x_1(t), \cdots, x_n(t))$, $t \in [a,b]$ is of class C^1 (equivalently, each x_j is of class C^1). Let us explore the notion "the length of C". Let P be the partition of $[a,b]$ defined by $a = t_0 < t_1 < \cdots < t_{\ell-1} < t_\ell = b$. Then

$$\sum_{i=0}^{\ell-1} \|g(t_{i+1}) - g(t_i)\| = \sum_{i=0}^{\ell-1}\left(\sum_{j=1}^n |x_j(t_{i+1}) - x_j(t_i)|^2\right)^{\frac{1}{2}}$$

$$= \sum_{i=0}^{\ell-1}\left(\sum_{j=1}^n \left|\frac{x_j(t_{i+1}) - x_j(t_i)}{t_{i+1} - t_i}\right|^2\right)^{\frac{1}{2}}(t_{i+1} - t_i) \tag{3.15}$$

can be considered to be an approximation to the "length of C". If we use the Mean Value Theorem we can express each summand of the interior sum in (3.15) as the quantity $|x_j(t_i + \theta_j^i(t_{i+1} - $

$t_i))|^2, 0 < \theta_j^i < 1$. Thus, (3.15) equals

$$\sum_{i=0}^{\ell-1} \left(\sum_{j=1}^{n} |x_j'(\tilde{t}_i^j)|^2 \right)^{\frac{1}{2}} (t_{i+1} - t_i), \tag{3.16}$$

where $\tilde{t}_i^j \in [t_i, t_{i+1}]$. It follows easily (since each x_j' is continuous), that, if $\|P\| \to 0$, the expressions in (3.15) (or (3.16)) tend to

$$\int_a^b \|g'(t)\| dt = \int_a^b \left(\sum_{j=1}^{n} \left(\frac{dx_j}{dt}(t) \right)^2 \right)^{\frac{1}{2}} dt. \tag{3.17}$$

Suppose $\varphi : [c,d] \xrightarrow{onto} [a,b]$ is a 1:1 differentiable function with $\varphi'(u) \neq 0$ for all u, then the image $g : [a,b] \to \mathbb{R}^n$ is the same as the image of $g \circ \varphi : [c,d] \to \mathbb{R}^n$. By the chain rule (or change of variables formula) we have

$$\int_a^b \|g'(t)\| dt = \int_c^d \|g'(\varphi(u))\| |\varphi'(u)| du = \int_c^d \|(g \circ \varphi)'(u)\| du. \tag{3.18}$$

This shows that what we considered to be a reasonable expression of the arc length of C is independent of the parametrization (as long as the two parametrizations g and $g \circ \varphi$ are related by the functions we introduced above). We are tempted to define a *smooth (differentiable)* curve C to be the equivalence class of all parametrizations that are related by such functions φ. Such a definition allows us to define the *arc length* of C to be the integral (3.17). There is another property of parametrizations that is often important: *orientation*. If $g : [a,b] \to \mathbb{R}^n$ is a parametrization of C, so is $g \circ \varphi$, where $\varphi(t) = b + a - t$ is a 1:1 onto function from $[a,b]$ onto $[a,b]$ having the property that $\varphi(a) = b$ and $\varphi(b) = a$. The parametrization $g \circ \varphi$ has the opposite orientation than the one involving g. As we have seen, arc length is independent of orientation. We shall see, however, that orientation plays an important role in various situations.

Suppose f is a continuous function whose domain includes a smooth curve C (i.e., the image of a parametrization of C). A natural extension of the integration we introduced in the last section is the *line integral of f over C*. We assume f is real-valued with a parametrization given by $x(t) = (x_1(t), \cdots, x_n(t)) = g(t), a \leq t \leq b$. We then define

$$\int_C f = \int_C f ds = \int_a^b f(g(t)) \|g'(t)\| dt. \tag{3.19}$$

The same argument we gave above in case $f \equiv 1$ can be applied to show that this integral is independent of the parametrization. We introduced ds to denote the element of arc length $\|g'(t)\| dt$. Let us examine this more carefully. Consider the function $s : [a,b] \to [0, L]$ defined by

$$s(t) = \int_a^t \|g'(u)\| du,$$

where $L = s(b)$ is the arc-length of C. Then $s'(t) = \|g'(t)\|$ for $t \in (a, b)$. Consider the inverse function $\varphi = s^{-1} : [0, L] \xrightarrow[\text{onto}]{1:1} [a, b]$. We know, since $\varphi(s(t)) = t$, that φ' exists and satisfies $\varphi'(s(t))s'(t) = 1$. Hence,

$$\varphi'(s(t)) = \frac{1}{\|g'(t)\|}, \tag{3.20}$$

or, letting $u = s(t)$,

$$\varphi'(u) = \frac{1}{\|g'(\varphi(u))\|} \qquad \text{since } t = \varphi(u)). \tag{3.20'}$$

Thus, if we let $\tilde{g} = g \circ \varphi$, then if $v \in (0, L)$, using (3.20'), Let s be replaced by u,

$$\int_0^v \|\tilde{g}'(u)\| ds = \int_0^v \|g'(\varphi(u))\| \cdot |\varphi'(u)| du \stackrel{(3.20')}{=\!=\!=} \int_0^v du = v,$$

$\tilde{g}$ is called *the arc-length parametrization* of C.

Of particular importance is the *scalar-valued line integral for vector fields*. A vector field is an $\mathbb{R}^n$-valued function defined on a region in $\mathbb{R}^n$. An example we have already encountered is the gradient, $\nabla f = (\frac{\partial f}{\partial x_1}, \frac{\partial f}{\partial x_2}, \cdots, \frac{\partial f}{\partial x_n})$ of an appropriately smooth function f. Let $\mathbf{F} = (f_1, f_2, \cdots, f_n)$ be a continuous vector-valued ($\mathbb{R}^n$-valued) function on an open set containing a curve C in $\mathbb{R}^n$ which we assume to be smooth (C is parametrized by a C^1 function $g : [a, b] \to \mathbb{R}^n$—the components of $\mathbf{g}$ are each continuous on $[a, b]$ and have continuous derivatives in (a, b)).[1] Then the line integral $\int_C \mathbf{F} \cdot d\mathbf{x}$ is defined by the integral

$$\int_a^b \mathbf{F}(\mathbf{g}(t)) \cdot \mathbf{g}'(t) dt, \tag{3.21}$$

where $\cdot$ denotes the inner product (of two vectors in $\mathbb{R}^n$). If we have chosen an orientation and consider only representations of C having the same orientation, the integral (3.21) is independent of the representation and we denote it by $\int_C \mathbf{F} \cdot d\mathbf{x}$ (as indicated above). This integral can be expressed as an integral of a scalar-value function on C (see (3.19)). Let $\mathbf{t}(\mathbf{g}(u)) = \frac{\mathbf{g}'(u)}{\|\mathbf{g}'(u)\|}$ be the *unit tangent vector*, $\mathbf{t}(\mathbf{x})$, *to the curve C at the point* $\mathbf{x}$ (in the "forward" or "chosen" direction). Then the integrand in (3.21) is $F_{tang}(\mathbf{x}) ds \equiv \mathbf{F}(\mathbf{g}(u)) \cdot \mathbf{t}(\mathbf{g}(u)) \|\mathbf{g}'(u)\| du$, where $F_{tang}(\mathbf{x})$ is the component of $\mathbf{F}(\mathbf{x})$ in the direction of $\mathbf{t}(\mathbf{x})$. Thus,

$$\int_C \mathbf{F} \cdot d\mathbf{x} = \int_C F_{tang} ds. \tag{3.22}$$

---

[1]We shall use the "bold face" notation to indicate that the function is n-dimensional vector-valued.

An important example of these integrals $\int_C \mathbf{F} \cdot d\mathbf{x}$ arises when $\mathbf{F}$ is a force field; in this case the integral represents the work done by the force on a particle that traverses the curve C.

(HW) 3.18   Find the arc length of the curve C parametrized by $\mathbf{g}(u) = (\log u, 2u, u^2)$, $u \in [1, e]$.

(HW) 3.19   Suppose $\mathbf{F}$ is a continuous map of a region in $\mathbb{R}^n$ containing a C^1 curve C into $\mathbb{R}^n$. Show that $|\int_C \mathbf{F} \cdot d\mathbf{x}| \leq \int_C \|\mathbf{F}\| ds$.

Suppose $\mathbf{F} = \nabla f$, where $f : U \subset \mathbb{R}^n \to \mathbb{R}$ with U open and $C \subset U$, where C is a C^1 curve. Suppose $\boldsymbol{\varphi} : [a, b] \to \mathbb{R}^n$ is a parametrization of C and $h = f \circ \boldsymbol{\varphi}$. Then $h'(t) = \nabla f(\boldsymbol{\varphi}(t)) \cdot \boldsymbol{\varphi}'(t)$ and it follows that $\int_a^b h'(t) dt = f(\boldsymbol{\varphi}(b)) - f(\boldsymbol{\varphi}(a))$ which tells us that

$$\int_C \mathbf{F} \cdot d\mathbf{x} = f(\boldsymbol{\varphi}(b)) - f(\boldsymbol{\varphi}(a)). \tag{3.23}$$

Let us apply the same argument when $\boldsymbol{\varphi}$ is replaced by $: [c, d] \to U$, a C^1 map whose range represents a curve $\tilde{C} \subset U$ such that $(c) = \boldsymbol{\varphi}(a)$ and $(d) = \boldsymbol{\varphi}(b)$. That is C and $\tilde{C}$ are two curves in U having the same end points $\mathbf{P}$ and $\mathbf{Q}$ in U (see Fig. 3.3).

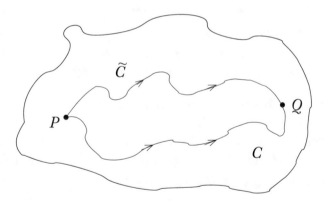

Figure 3.3: The two curves of C and $\tilde{C}$.

Then

$$\int_C \mathbf{F} \cdot d\mathbf{x} = f(\mathbf{Q}) - f(\mathbf{P}) = \int_{\tilde{C}} \mathbf{F} \cdot d\mathbf{x}. \tag{3.24}$$

This shows that the line integral of the vector field $\mathbf{F}$ on U is independent of the curve $C \subset U$ as long as the end points $\mathbf{P}$ and $\mathbf{Q}$ do not change and $\nabla f = \mathbf{F}$.

This independence of the end points is not valid for line integrals of *all* C^1 vector fields. Consider the following example: Let $\mathbf{F}(x, y) = (2xy, x^2 - y^2)$ and $\tilde{\mathbf{F}}(x, y) = (2xy, y^2 - x^2)$. $\mathbf{F}$ is the gradient of $f(x, y) = yx^2 - \dfrac{y^3}{3}$. Let C be the curve with representation $\varphi(t) = (t, t)$, $0 \le t \le 1$, and $\tilde{C}$ the curve with representation $(t) = (t, t^2)$, $0 \le t \le 1$. C and $\tilde{C}$ have the same end points $(0, 0)$ and $(1, 1)$. Thus, by (3.24),

$$\int_C \mathbf{F} \cdot d\mathbf{x} = f(1, 1) - f(0, 0) = \frac{2}{3} = \int_{\tilde{C}} \mathbf{F} \cdot d\mathbf{x}. \tag{3.25}$$

On the other hand,

$$\int_C \tilde{\mathbf{F}} \cdot d\mathbf{x} = \frac{2}{3} \ne \frac{1}{3} = \int_{\tilde{C}} \tilde{\mathbf{F}} \cdot d\mathbf{x}. \tag{3.26}$$

(HW) 3.20   Carry out the computations that justify the various assertions made in the above paragraph that contains (3.25) and (3.26).

The fact that $\mathbf{F} = \nabla f$ satisfies (3.24) is often stated by saying $\mathbf{F}$ is a conservative vector field. That is, $\mathbf{F}$ is a *conservative* vector field if it satisfies

(a) if C_1 and C_2 are any two oriented curves in U with the same end points, then $\int_{C_1} \mathbf{F} \cdot d\mathbf{x} = \int_{C_2} \mathbf{F} \cdot d\mathbf{x}$. This is equivalent to

(b) If C is a closed curve in U, then $\int_C \mathbf{F} \cdot d\mathbf{x} = 0$ ("closed" means that the starting and ending points of C are the same).

We will assume, in this discussion, that the domain, U, of $\mathbf{F}$ is a connected open set. This means that any two points in U can be connected by a C^1 curve.

Theorem 3.21   *A continuous vector field $\mathbf{F}$ on a connected open set U is conservative iff $\mathbf{F}$ is the gradient of a C^1 function on U.*

Proof: We have just shown that a gradient is a conservative field. Thus, suppose $\mathbf{F}$ is conservative in U. Fix $\vec{y}_0 \in U$ and consider another point $\vec{y} \in U$. Let C be a C^1 curve in U from $\vec{y}_0$ to $\vec{y}$ (by hypothesis such a curve always exists) and define $f(\vec{y})$ to be $\int_C \vec{F} \cdot d\vec{x}$. Then f is a well-defined function on U since it is not dependent on C (as long as it starts at $\vec{y}_0$ and ends at $\vec{y}$) because $\mathbf{F}$ is conservative. (We are using $\vec{y}$ and $\vec{F}$ for $\mathbf{y}$ and $\mathbf{F}$.) We shall show

$$\mathbf{F}(\vec{y}) = (\frac{\partial f}{\partial y_1}(\vec{y}), \cdots, \frac{\partial f}{\partial y_n}(\vec{y})) = \nabla f(\vec{y}).$$

Let $\vec{h} = (h, 0, \cdots, 0)$ with h sufficently small so that the line segment from $\vec{y}$ to $\vec{y} + \vec{h}$ lies with U. Let L denote this line segment (which we regard as a curve from $\vec{y}$ to $\vec{h} + \vec{y}$). We can adjoin it to C to form a curve from $\vec{y}_0$ to $\vec{y}_0 + \vec{h}$ (convince yourself that the various arguments we have used apply to such piecewise smooth curves). Having done so, we have

$$\frac{f(\vec{y} + \vec{h}) - f(\vec{y})}{h} = \frac{1}{h} \int_L \mathbf{F} \cdot d\mathbf{x} = \frac{1}{h} \int_0^h F_1(y_1 + t, y_2, \cdots, y_n) dt$$

and the last term tends to $\dfrac{\partial f}{\partial y_1}(\vec{y}) = F_1(y)$. Applying this argument to the other coordinates of $\vec{y}$ we obtain the desired result. ∎

The function so obtained, f, is called the *potential* of the conservative field. It is unique up to an additive constant.

3.3 HIGHER DIMENSIONAL INTEGRALS

Much of the basic theory of integrals of real valued functions on n-dimensional regions is practically the same as the two-dimensional case. Since the latter is notationally simpler and geometrically easier to visualize, we begin with the two-dimensional case.

Let $R = [a, b] \times [c, d] = \{(x, y) \in \mathbb{R}^2 : x \in [a, b], y \in [c, d]\}$ be a *rectangle* in $\mathbb{R}^2$. A *partition*, P, of R is obtained from a partition $a = x_0 < x_1 < \cdots < x_M = b$ of $[a, b]$ and another partition $c = y_0 < y_1 < \cdots < y_N = d$ of $[c, d]$ by forming the subrectangles $R_{mn} = [x_{m-1}, x_m] \times [y_{n-1}, y_n]$, $0 \le m \le M$, $0 \le n \le N$. Clearly, $R = \displaystyle\bigcup_{m,n=0}^{M,N} R_{mn}$ and each R_{mn} has area $\triangle_{mn} = [x_m - x_{m-1}][y_n - y_{n-1}]$. Let f be a bounded real valued function defined on R. If $m_{jk} = \inf\{f(x, y) : (x, y) \in R_{jk}\}$ and $M_{jk} = \sup\{f(x, y) : (x, y) \in R_{jk}\}$ we let the *lower* and *upper (Riemann) sums* of f, associated with P, be

$$\ell(P) = \sum_{j=1}^M \sum_{k=1}^N m_{jk} \triangle_{jk} \text{ and } u(P) = \sum_{j=1}^M \sum_{k=1}^N M_{jk} \triangle_{jk}.$$

As in the one-dimensional case,

$$\underline{\int_R} f = \sup_P \ell(P) \text{ and } \overline{\int_R} f = \inf_P u(P) \tag{3.27}$$

are the *lower* and *upper* integrals of f on R. If the two quantities in (3.27) are equal, f is said to be *(Riemann) integrable* and this common value is called the *(Riemann) integral* of f over R and is denoted by

$$\int_R f \text{ or } \int_R f(x, y) dx dy.$$

All this is very much like the one-dimensional case.

The first significant difference arises when we want to integrate an f defined on a "reasonable" connected region that is not a rectangle (such as a closed disk or a bounded closed region that is more general). Suppose S is such a bounded region in $\mathbb{R}^2$. Consider the *characteristic (or indicator) function* of S

$$\chi_S(x, y) = \begin{cases} 1 & \text{if } (x, y) \in S \\ 0 & \text{if } (x, y) \notin S \end{cases} \tag{3.28}$$

(actually, the definition (3.28) applies to any $S \subset \mathbb{R}^2$). Let R be a large enough rectangle so that $S \subset R$. We can then say that f *integrable on* S iff $\chi_S f$ is integrable on R (extend f to be 0 outside its domain S). In this case we define the integral of f over S to be $\int_R \chi_S f \equiv \int_S f$.

Several basic properties of this two-dimensional integral are proved by the same arguments that gave us Theorem 3.10, (3.7), (3.8), (3.9), (3.10), (3.11), and (3.12). Here are some of these:

Theorem 3.22

(a) If f and g are integrable on a bounded set $S \subset \mathbb{R}^2$ and $a, b \in \mathbb{R}$, then $af + bg$ is integrable on S and

$$\int_S af + bg = a \int_S f + b \int_S g.$$

(b) If f and g are integrable on S and $f \leq g$, then $\int_S f \leq \int_S g$.

(c) If S_1 and S_2 are two bounded sets in $\mathbb{R}^2$ and f is a bounded function on $S_1 \cup S_2$, then f is integrable on S_1 and on S_2 iff it is integrable on $S_1 \cup S_2$ and, if this is true and $S_1 \cap S_2 = \emptyset$,

$$\int_{S_1 \cup S_2} f = \int_{S_1} f + \int_{S_2} f.$$

(d) If f is integrable on a bounded set S, then $|f|$ is integrable on S and $|\int_S f| \leq \int_S |f|$.

As in the one-dimensional case, it is natural to ask what conditions of f guarantee integrability. In Section 3.1 we found that continuity on $[a, b]$ is a natural condition. Continuity on a rectangle presents us a problem when we want to use the definition of $\int_S f$ we gave above. If f is continuous on $S \subset R$ (rectangle), the function $\chi_S f$ we considered there, in general, will not be continuous on R. The notion of *content zero* will help us show that, here again, continuity plays an important role in giving a condition for integrability.

Definition 3.23   *A set $B \subset \mathbb{R}^2$ has content zero iff for each $\epsilon > 0$ there exists a finite collection of rectangles $R_1, R_2, \cdots, R_L$ such that $B \subset \bigcup_{\ell=1}^{L} R_\ell$ and the sum of the areas of the R_ℓ's is less than ϵ.*

Theorem 3.24

(i) If $B \subset \mathbb{R}^2$ has content zero so does any subset of B;

(ii) A finite union of sets of content zero has content zero;

(iii) Suppose $g : (a_0, b_0) \to \mathbb{R}^2$ is of class C^1, then $g([a, b])$ has content zero if $a_0 < a < b < b_0$.

(HW) 3.25   Prove Theorem 3.24.

Here is the condition of integrability we are seeking:

Theorem 3.26   *Let f be a bounded function on a rectangle R. If the set of points in R for which f is not continuous has content zero, then f is integrable in R.*

(HW) 3.27   Prove Theorem 3.26.

It is clear from this last theorem that if f is a bounded function that is continuous on a bounded set S that has a boundary, ∂S, of content zero, then f is integrable on S. This is clear from the fact that χ_S is continuous at all points of $\mathbb{R}^2$ that are not in ∂S.

Most domains, S, of functions that we shall encounter have boundaries that are images of parametrizations of C^1 (closed) curves. From part (iii) of Theorem 3.24 it follows that these boundaries have content zero.

It makes sense, therefore, that we can assign a notion of *area* for each bounded set S; such set ∂S has content 0: that is, the area of S is $\int_S 1 = \int_S \chi_S$. The domains we described in the paragraph before this one, therefore, have an area $A(S) = \int_S 1$.

One has to be careful about this notion, however. Let us consider the following example. This will be an open subset, S, of the unit square $[0, 1] \times [0, 1]$. Consider the points

$$\frac{1}{2}, \frac{1}{4}, \frac{3}{4}, \frac{1}{8}, \frac{3}{8}, \frac{5}{8}, \frac{7}{8}, \quad \cdots \quad , \frac{1}{2^k}, \frac{3}{2^k}, \frac{5}{2^k}, \quad \cdots \quad, \frac{2^k - 1}{2^k}, \cdots$$

in $(0, 1)$. We construct the following open rectangles: each has the form $I \times (0, 1)$, where on the x-axis they are ordered so that the first is of length $\frac{1}{4}$ centered at $\frac{1}{2}$, the next two are of length $\frac{1}{16}$ with centers at $\frac{1}{4}$ and $\frac{3}{4}$, the next four are of length $\frac{1}{64}$, centered at $\frac{1}{8}, \frac{3}{8}, \frac{5}{8}$ and $\frac{7}{8}$, and so on. The set S is the union of these open rectangles. Since $\sum_{j=2}^{\infty} 2^{-j} = \frac{1}{2}$, S is the union of countably many open rectangles of "total area" $\leq \dfrac{1}{2}$ (that is, the area of each finite union of these rectangles is $< \dfrac{1}{2}$ and this is a disjoint union).

(HW) 3.28   Show that S is dense in $R = [0, 1] \times [0, 1]$. Since S is open, $R - S$ is closed and, in fact, it is the boundary, ∂S, of S.

(HW) 3.29   Show that $\displaystyle\overline{\int}_R \chi_S = 1$ and $\displaystyle\underline{\int}_R \chi_S = 0$.

This is an example of an open, bounded set $S \subset \mathbb{R}^2$ such that χ_S is *not* integrable.

The extension of this theory of integration to n dimensions, $n \geq 3$, is straight forward. The two-dimensional rectangles are replaced by n-dimensional rectangles

$$
\begin{aligned}
R \;\; &= [\alpha_1, \beta_1] \times \cdots \times [\alpha_n, \beta_n] \\
&= \{x \in \mathbb{R}^n : \alpha_1 \leq x_1 \leq \beta_1, \cdots, \alpha_n \leq x_n \leq \beta_n\}
\end{aligned}
$$

The n-dimensional volume of R is $\displaystyle\prod_{j=1}^{n} (\beta_j - \alpha_j)$. A partition of R is obtained by partitioning each of its "sides" $[\alpha_j, \beta_j]$, $j = 1, 2, \cdots, n$. The notion of content zero for a bounded set $S \subset \mathbb{R}^n$ is that, for each $\epsilon > 0$, there are a finite numbers of (n-dimensional) rectangles, $R_1, R_2, \cdots, R_L$, such that $S \subset \displaystyle\bigcup_{j=1}^{L} R_j$ and the sum of their volumes is less than ϵ. C^1 curves are extended to C^1 *submanifolds* that are given parametrically by C^1 maps $f : [a_1, b_1] \times \cdots \times [a_m, b_m] \to \mathbb{R}^n$, $m < n$, with a notion of equivalence that is the obvious extension of the one we introduced for curves. We will give more details about these integrals at appropriate times in the sequel. For the moment just convince yourselves that these extensions are plausible and are not very different from the two-dimensional situation.

3.4   MULTIPLE INTEGRALS AND THEIR REDUCTION TO ONE DIMENSIONAL INTEGRALS

We will show that the integration we described on regions in $\mathbb{R}^n$ can be performed by iterating one-dimensional integrals. This is an important fact that has many applications and is most useful for the calculation of the n-dimensional integrals. Again, we begin with $n = 2$.

Let f be a continuous function on the finite rectangle $R = [a, b] \times [c, d] \subset \mathbb{R}^2$. Fix $x \in [a, b]$, then $f(x, y)$ is a continuous function of y on $[c, d]$. Thus, the integral

$$\varphi(x) = \int_c^b f(x, y) dy \tag{3.29}$$

is well defined. Furthermore it provides us with a function φ on $[a, b]$. Since f must be uniformly continuous (Theorem 1.46 in Chapter 1) it follows easily that φ is continuous. Thus

$$\int_a^b \varphi(x) dx = \int_a^b \left(\int_c^d f(x, y) dy \right) dx \equiv \int_a^b \int_c^d f(x, y) dy dx \tag{3.30}$$

is defined.

We can, of course, interchange the role of the two variables and obtain

$$\int_c^d \psi(y) dy \equiv \int_c^d \left(\int_a^b f(x, y) dx \right) dy \equiv \int_c^d \int_a^b f(x, y) dx dy. \tag{3.31}$$

We claim that these two "iterated" integrals are equal and, in fact, each is equal to $\int_R f$. This is easy to see. Here is a sketch of a proof. Let $P = P_{JK}$ be a partition of R obtained by partitioning $[a, b]$ and $[c, d]$ into J and K equal subintervals of lengths $\Delta x = \dfrac{b - a}{J}$ and $\Delta y = \dfrac{d - c}{K}$. Observe that

$$\sum_{k=1}^{K} \left(\sum_{j=1}^{J} f(x_j, y_k) \Delta x \right) \Delta y = \sum_{\substack{1 \leq k \leq K \\ 1 \leq j \leq J}} f(x_j, y_k) \Delta x \Delta y$$

$$= \sum_{j=1}^{J} \left(\sum_{k=1}^{K} f(x_j, y_k) \Delta y \right) \Delta x.$$

Using uniform continuity we see that the middle term is between $\ell(P)$ and $u(P)$ and these two numbers are as close to each other as we wish as J and K tend to ∞. Thus, each of these expressions tend to $\int_R f$. Then letting $J \to \infty$ in the first sum and, after this, letting $K \to \infty$, we see that, in the limit we obtain (3.31). The same observation applied to the third sum shows that, in the limit, we obtain (3.30).

If $f(x, y)$ is a polynomial we can easily obtain $\varphi(x)$, which is, again, a polynomial (in x). Thus we can evaluate $\int_R f$ explicitly. Show, for example, that, if $R = [1, 2] \times [2, 3]$ and $f(x, y) = x^2 + 2y^2 - xy + 1$, then $\int_R f = \dfrac{49}{4}$. What is this integral if $R = [2, 3] \times [1, 2]$?

(HW) 3.30   Calculate $\int_R f(x, y)dxdy$ when

(a) $R = [0, 1] \times [0, 1]$ and $f(x, y) = xe^{x+y}$,

(b) $R = [1, 2] \times [1, 2]$ and $f(x, y) = \dfrac{x}{x + y}$.

It is not hard to extend this equality to the more general case of an integrable function f on R provided, for each $y \in [c, d]$, the function f_y, having values $f_y(x) = f(x, y)$, is integrable on $[a, b]$, and $g(y) = \int_a^b f(x, y)dx$ is integrable on $[c, d]$. In this case,

$$\int_R f = \int_c^d \left[\int_a^b f(x, y)dx \right] dy.$$

It is not hard to extend these observations to rectangles in $\mathbb{R}^n$. Convince yourself of this fact.

What we presented here, the fact that for these functions f, we have

$$\int_R f = \int_c^d \left(\int_a^b f(x, y)dx \right) dy = \int_a^b \left(\int_c^d f(x, y)dy \right) dx, \tag{3.32}$$

is known as *Fubini's theorem* (for the rectangle).

We can extend this result to other regions besides rectangles. For example, suppose

$$S = \{(x, y) : a \leq x \leq b, \varphi(x) \leq y \leq \psi(x)\}$$

See Fig. 3.4.

Assume φ and ψ are smooth so that $\chi_S f$ is integrable over $R = [a, b] \times [c, d]$, $c \leq \inf_{x\in[a,b]} \varphi(x)$ and $d \geq \sup_{x\in[a,b]} \psi(x)$. It follows that

$$\int_R \chi_S f = \int_S f = \int_a^b \left(\int_{\varphi(x)}^{\psi(x)} f(x, y)dy \right) dx. \tag{3.33}$$

(HW) 3.31   Calculate $\int_S f$ when $f(x, y) = x + y$ and S is the region between the two curves represented by the function $\varphi(x) = 2x^3$ and $\psi(x) = 2\sqrt{x}$. (N.B. the interval $[a, b]$ is determined by these two functions—also, we assume $x \geq 0$.)

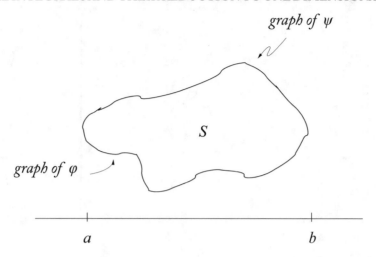

graph of ψ

S

graph of φ

a b

Figure 3.4: The graph of S.

(HW) 3.32   Can you show that $\displaystyle\int_S f$ can also be calculated as an integral of the form

$$\int_c^d \left(\int_{\mu(y)}^{\nu(y)} f(x, y)dx \right) dy?$$

That is, find μ, ν, c, and d. Show that you obtain the same value as in (HW) 3.31.

Such regions are shown in Fig. 3.5.
We will call these *standard regions*. Their boundaries are made up of straight line segments and curves that are functions of x or functions of y. A disk can be represented as a region whose boundary is made up of two functions of x or two functions of y. Figure 3.6 explains this.

Observe that the boundary does not contain any line segment that contains more than one point. One can break up some more general regions into a finite number of standard regions; thus, Fubini's theorem can be used for evaluating integrals over these more general regions.

As indicated in Section 3.3, the Riemann integral can be extended to higher dimensions in a natural way. In particular, Fubini's theorem for functions that are continuous on finite rectangles extends immediately from the two-dimensional case. A bit more challenging is the extension of standard regions. Let us consider the three-dimensional case. Suppose T is an elementary region in the xy-plane and γ_1, γ_2 are two functions on T such that $\gamma_1(x, y) \le \gamma_2(x, y)$ for $(x, y) \in T$, then

$$S = \{(x, y, z) \in \mathbb{R}^3 : (x, y) \in T \text{ and } \gamma_1(x, y) \le z \le \gamma_2(x, y)\}$$

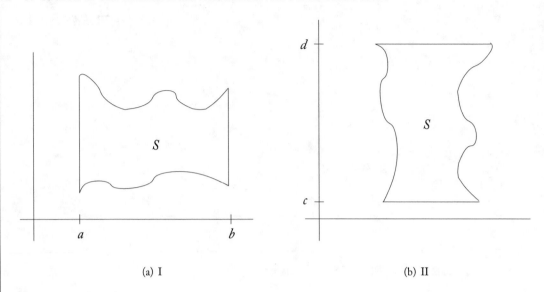

(a) I   (b) II

Figure 3.5: Standard regions.

is an elementary regioin in $\mathbb{R}^3$. Suppose

$$T = \{(x, y) \in \mathbb{R}^2 : a \leq x \leq b, \varphi_1(x) \leq y \leq \varphi_2(x)\}.$$

Then, it is rather straight-forward to show that the integral of a continuous f on S is integrable and

$$\int_S f = \int_a^b \left[\int_{\varphi_1(x)}^{\varphi_2(x)} \left(\int_{\gamma_1(x,y)}^{\gamma_2(x,y)} f(x, y, z) dz \right) dy \right] dx. \tag{3.34}$$

Similarly,

$$\int_S f = \int_c^d \left[\int_{\psi_1(y)}^{\psi_2(y)} \left(\int_{\gamma_1(x,y)}^{\gamma_2(x,y)} f(x, y, z) dz \right) dx \right] dy \tag{3.35}$$

if $T = \{(x, y) \in \mathbb{R}^2 : c \leq y \leq d, \psi_1(y) \leq x \leq \psi_2(y)\}$.

(HW) 3.33   Let S be the region bounded by $z = 0, z = \pi, y = 0, y = 1, x = 0$ and $x + y = 1$. Compute $\int_S f$ when $f(x, y, z) = x^2 \cos z$.

3.5   GREEN'S THEOREM

Suppose S is a bounded region in $\mathbb{R}^2$ that has a boundary that is closed curve. We begin with $S = R$, a rectangle, $[a, b] \times [c, d]$. The boundary, ∂R, of R consists of the four line segments C_1, C_2, D_1

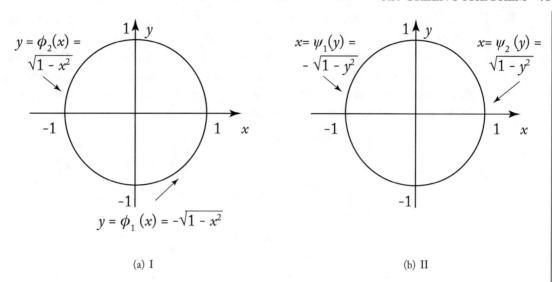

(a) I (b) II

Figure 3.6: Disks.

and D_2 (see Fig. 3.7). We assume that $\partial R = C$ is a curve oriented in the *positive direction*. By

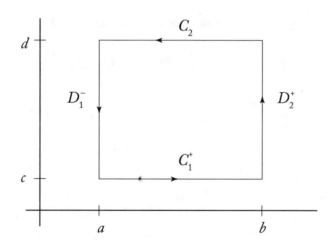

Figure 3.7: A region with closed curve boundary.

this we mean that as we move along the boundary ∂R we keep the interior of the domain R on our left. Let $\vec{\varphi_1} : [a, b] \to \mathbb{R}$ be the parametrization of C_1 satisfying $\vec{\varphi_1}(x) = (x, c)$; $\vec{\varphi_2}(x) = (x, d)$, $a \leq x \leq b$, $\vec{\psi_1}(y) = (a, y)$, $c \leq y \leq d$ and $\vec{\psi_2}(y) = (b, y)$, $c \leq y \leq d$, are, respectively, parametrizations of C_2, D_1 and D_2. Consistent with the notion of orientation we introduced on page 61, we claim that the curve $\partial R = C^+$ equals $C_1^+ + D_2^+ + C_2^- + D_1^-$ with $+$ indicating the positive and $-$ the negative orientaion; we consider C^+ as made up by the four curves in the sum, the first followed by the second which is followed by the third and D_1^- is the last one.

Let $\overrightarrow{F} = (P, Q)$ be a vector field defined in a domain that includes the closed rectangle R. Let us assume that it is C^1 (a bit more than we need). Then, using Fubini's theorem and the fundamental theorem of calculus, we have

$$\iint_R \left(-\frac{\partial P}{\partial y} + \frac{\partial Q}{\partial x}\right) dxdy = \int_a^b \left(\int_c^d \left(-\frac{\partial P}{\partial y}(x, y)dy\right)\right) dx +$$

$$\int_c^d \left(\int_a^b \frac{\partial Q}{\partial x}(x, y)dx\right) dy = \int_a^b \left(-(P(x, d) - P(x, c))\right) dx +$$

$$\int_c^d (Q(b, y) - Q(a, y))dy. \tag{3.36}$$

We shall prove the following

Lemma 3.34

$$\iint_R \left(-\frac{\partial P}{\partial y} + \frac{\partial Q}{\partial x}\right) dxdy = \int_{C+} \overrightarrow{F} \cdot d\overrightarrow{u}. \tag{3.37}$$

Proof: Making use of the parametrizations we just introduced we have

$$\int_{C_1^+} \overrightarrow{F} \cdot d\overrightarrow{u} = \int_a^b (P(x, c), Q(x, c)) \cdot (1, 0)dx = \int_a^b P(x, c)dx,$$

$$\int_{D_2^+} \overrightarrow{F} \cdot d\overrightarrow{u} = \int_c^d (P(b, y), Q(b, y)) \cdot (0, 1)dy = \int_c^d Q(b, y)dy,$$

$$\int_{C_2^-} \overrightarrow{F} \cdot d\overrightarrow{u} = \int_a^b (P(a + b - x, d), Q(a + b - x, d)) \cdot (-1, 0)dx$$

$$= \int_a^b (-P(a + b - x, d)) \, dx = \int_a^b (-P(x, d)) \, dx$$

and, finally,

$$\int_{D_1^-} \overrightarrow{F} \cdot d\overrightarrow{u} = \int_c^d (-P(a, c + d - y), Q(a, c + d - y)) \cdot (0, -1)) \, dy$$

$$= -\int_c^d Q(a, c + d - y)dy = -\int_c^d Q(a, y)dy.$$

We are using the result in this exercise:

(HW) 3.35   Show that if f is a continuous function on $[c, d]$, then

$$\int_c^d f(c + d - y)dy = \int_c^d f(y)dy.$$

We are also using the representations $x \to a + b - x$ and $y = c + d - y$ for the C_2^- and D_1^- integration.

Equality (3.37) is now an immediate consequence of (3.36) and the four equalities we obtained.   ∎

This result extends to very general regions S whose boundary ∂S is a C^1 curve. We first observe that if R_1 and R_2 are two rectangles that have a portion of a single side in common and disjoint interiors, then (3.37) is true for $S = R_1 \bigcup R_2$. Let us be a bit more specific. Consider the case in Fig. 3.8.

The arrows indicate the positive orientation of ∂R_1, ∂R_2 and ∂S. Along the line segment containing the line joining the point (b, α) to the point (b, β), the line integrals involved in ∂R_1 and ∂R_2 cancel each other when they traverse the segment joining (b, α) to (b, β). Thus, we obtain the extension of (3.37):

$$\iint_S \left(-\frac{\partial P}{\partial y} + \frac{\partial Q}{\partial x} \right) dxdy = \int_{\partial S} \vec{F} \cdot d\vec{u} \tag{3.38}$$

for P, Q of class C^1 in any domain that contains $\overline{S}$ with ∂S the boundary curve considered in the positive direction.

Much more general regions can be considered for which (3.38) holds and this fact is obtained by extending this method. Let S be a bounded domain with a C^1 boundary ∂S. For a "fine" grid (as in the accompanying figure), form the (finite) union of rectangles that approximates S and is interior to S. Then ∂R approximates ∂S. By a finite number of arguments that produced (3.38) we see that if P and Q are C^1 functions in a region that contains S, then we see that (3.38) is valid for R. Then, by an approximating argument that should be quite plausible to you, we obtain

Theorem 3.36   Green's theorem–our version.   *Let U be an open, bounded domain that includes S and its boundary ∂S (of class C^1). Then (3.38) is true whenever (P, Q) is a C^1 vector field in the larger domain U.*

In most textbooks the vector field is of the form (P, Q) and the equality of Green's theorem is written in the form:

$$\int_{\partial S} Pdx + Qdy = \iint_X \left(\frac{\partial Q}{\partial x} - \frac{\partial P}{\partial y} \right) dxdy. \tag{3.39}$$

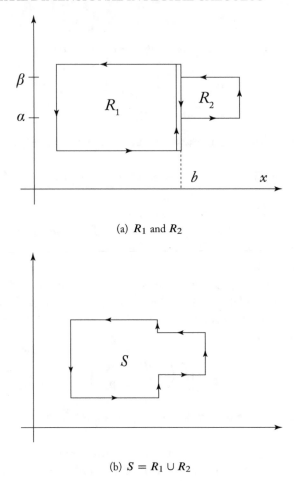

(a) R_1 and R_2

(b) $S = R_1 \cup R_2$

Figure 3.8: Regions.

The form we presented in (3.38) explains what is meant by the integral on the left in (3.39). Figure 3.7 indicates a further generality of the domains that can be considered.

Corollary 3.37   *If S is a region bounded by a C^1 curve (of the type we have described), then the area of S is*

$$A = \frac{1}{2} \int_{\partial S} x\,dy - y\,dx. \tag{3.40}$$

(HW) 3.38   Prove Corollary 3.37.

There are other important ways of writing Green's Theorem. We shall explore some of these. First of all, since P and Q are independent components of a vector field, we can set either of these functions to be 0. The area formula (3.40) has, thus, the variants

$$A = \int_{\partial S} x\, dy = -\int_{\partial S} y\, dx. \tag{3.40'}$$

Let R_+ and R_- be the counterclockwise and clockwise rotations (in $\mathbb{R}^2$) by $90°$. Thus, $R_+(x, y) = (-y, x)$ and $R_-(x, y) = (y, -x)$. If $\overrightarrow{t} = \overrightarrow{t}(s) = (t_1(s), t_2(s)) = (t_1, t_2)$ is the unit tangent to ∂S at $\overrightarrow{u}(s)$ (the arc length parametrization of ∂S), so that $\overrightarrow{t}$ points in the forward direction, then $\overrightarrow{n} = R_-\overrightarrow{t} = (t_2, -t_1)$ is the unit normal to ∂S directed outwardly form S. Suppose $\overrightarrow{F} = (F_1, F_2) = (P, Q)$. Then $\overrightarrow{F} \cdot \overrightarrow{n} = F_1 t_2 - F_2 t_1 = (-F_2, F_1) \cdot (t_1, t_2) = \overrightarrow{\tilde{F}} \cdot \overrightarrow{t}$, where $\overrightarrow{\tilde{F}} = (-F_2, F_1) = R_+\overrightarrow{F}$. Assuming the hypotheses of Green's Theorem (Theorem 3.36) we can apply it to $\overrightarrow{\tilde{F}}$ and we see that

$$\int_{\partial S} \overrightarrow{F} \cdot \overrightarrow{n}\, ds = \int_{\partial S} \overrightarrow{\tilde{F}} \cdot \overrightarrow{t}\, ds = \iint_S \left(\frac{\partial Q}{\partial y} + \frac{\partial P}{\partial x} \right) dx\, dy = \iint_S \left(\frac{\partial F_1}{\partial x} + \frac{\partial F_2}{\partial y} \right) dx\, dy.$$

If $\overrightarrow{F} = (F_1, F_2, \cdots, F_n)$ is a vector field defined on a domain $U \subset \mathbb{R}^n$, then the *divergence* of $\overrightarrow{F}$, $div\overrightarrow{F}$ is the function $\frac{\partial F_1}{\partial x_1} + \frac{\partial F_2}{\partial x_2} + \cdots + \frac{\partial F_n}{\partial x_n}$) (we assume, of course, that the partial derivatives exist). We have just shown, under the hypotheses of Green's theorem,

Theorem 3.39   The Divergence Theorem in $\mathbb{R}^2$.

$$\int_{\partial S} \overrightarrow{F} \cdot \overrightarrow{n}\, ds = \iint_S (div\overrightarrow{F})dx\, dy. \tag{3.41}$$

Recall (Theorem 2.39) that asserted that a continuous vector field is conservative iff it is the gradient of a C^1 function. This is corroborated by Green's theorem. If we add the hypothesis that $\overrightarrow{F} = \nabla f$, f a C^1 function on U, then, by (3.38),

$$\int_{\partial S} \overrightarrow{F} \cdot \overrightarrow{t}\, ds = \iint_S \left(\frac{\partial}{\partial x}\frac{\partial f}{\partial y} - \frac{\partial}{\partial y}\frac{\partial f}{\partial x} \right) dx\, dy = 0. \tag{3.42}$$

Another interesting equality involving a gradient vector field $\overrightarrow{F} = \nabla f$ is obtained by using (3.41): $\nabla f \cdot \overrightarrow{n}$ is called the *normal derivative* of f on ∂S; it is often denoted by $\frac{\partial f}{\partial n}$. Applying (3.41) we obtain

$$\int_{\partial S} \frac{\partial f}{\partial n} ds = \iint_S div\nabla f\, dx\, dy = \iint_S \triangle f\, dx\, dy. \tag{3.43}$$

Observe that $div \nabla = \dfrac{\partial^2}{\partial x^2} + \dfrac{\partial^2}{\partial y^2} = \triangle$ is the Laplace operator.

3.6   INTEGRATION ON SURFACES

A smooth surface S should be thought of the "natural" extension of a curve. We will study these surfaces in $\mathbb{R}^3$. We first consider such a surface S to be represented parametrically as the image of an open set, W, that is open and bounded in $\mathbb{R}^2$ by a $1:1$ C^1 map $\vec{g}: W \to \mathbb{R}^3: \vec{x} = (x, y, z) = \vec{g}(u, v)$, $(u, v) \in W$. The more general surface we consider is one that consists of finitely many (disjoint) parts of the type we just described. We have seen the analogous situation with curves. Equivalent parametrizations are defined as they were in the case of curves. As is the case for curves, *orientation* of a smooth surface is an important notion. Intuitively, orientation determines which side of a surface is positive or negative. This notion of a surface having "two sides" is associated with a choice of a unit normal vector that varies continuously as a point on the surface varies. If we have a "closed" surface, such as the surface of a ball in $\mathbb{R}^3$, this notion is, apparently, clear. Unfortunately, not every surface can be oriented. You are probably acquainted with a *Möbius strip* (see Fig. 3.9) which can be built by taking a long strip of paper, giving it a half-twist, and gluing together the two ends.

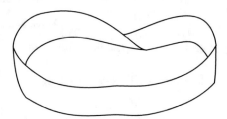

Figure 3.9: Möbius strip.

In order to study orientable surfaces we will use the operation of the *vector product* or *cross product* of two vectors $\vec{a} = (a_1, a_2, a_3)$ and $\vec{b} = (b_1, b_2, b_3)$ in $\mathbb{R}^3$:

$$\vec{a} \times \vec{b} \equiv (a_2 b_3 - a_3 b_2, a_3 b_1 - a_1 b_3, a_1 b_2 - a_2 b_1). \tag{3.44}$$

There are other ways of representing the cross product. The determinant of a 3×3 matrix

$$\begin{pmatrix} a_{11} & a_{12} & a_{13} \\ a_{21} & a_{22} & a_{23} \\ a_{31} & a_{32} & a_{33} \end{pmatrix}$$

having coefficients in $\mathbb{R}$ is the quantity

$$\Delta \equiv \begin{vmatrix} a_{11} & a_{12} & a_{13} \\ a_{21} & a_{22} & a_{23} \\ a_{31} & a_{32} & a_{33} \end{vmatrix} = a_{11} \begin{vmatrix} a_{22} & a_{23} \\ a_{32} & a_{33} \end{vmatrix} - a_{12} \begin{vmatrix} a_{21} & a_{23} \\ a_{31} & a_{33} \end{vmatrix} + a_{13} \begin{vmatrix} a_{21} & a_{22} \\ a_{31} & a_{32} \end{vmatrix},$$

where $\begin{vmatrix} a & b \\ c & d \end{vmatrix} = ad - bc$ is the determinant of the 2×2 matrix $\begin{pmatrix} a & b \\ c & d \end{pmatrix}$. The symbols $\vec{i}, \vec{j}, \vec{k}$ are often used to denote the canonical orthonormal vectors $\vec{e^1} = (1, 0, 0), \vec{e^2} = (0, 1, 0)$ and $\vec{e^3} = (0, 0, 1)$. The vector product, as defined in (3.44), can, symbolically, be represented by the "determinant"

$$\begin{vmatrix} \vec{i} & \vec{j} & \vec{k} \\ a_1 & a_2 & a_3 \\ b_1 & b_2 & b_3 \end{vmatrix}. \tag{3.45}$$

(HW) 3.40 (a) Show that $\vec{a} \times \vec{b}$ is orthogonal to $\vec{a}$ and $\vec{b}$ (and, thus, $\vec{a} \times \vec{b}$ is perpendicular to the plane generated by $\vec{a}$ and $\vec{b}$ when these two vectors are linearly independent);

(b) Let θ be the angle between $\vec{a}$ and $\vec{b}, 0 \leq \theta \leq \pi$. Show that $\|\vec{a} \times \vec{b}\| = \|\vec{a}\| \|\vec{b}\| \sin \theta$, the area of the parallelogram spanned by $\vec{a}$ and $\vec{b}$.

Let us now pass to the subject of integration on surfaces. We begin with the notion of surface area. Let us go back to the parametrization of S the map $\vec{g} : W \to \mathbb{R}^3$ we introduced at the beginning of this section. Consider a small rectangle in the open set W with vertices $(u, v), (u + \Delta u, v), (u, v + \Delta v)$ and $(u + \Delta u, v + \Delta v)$. Its image under $\vec{g}$ is a small "quadrilateral" (with curved sides) on the surface S having vertices $\vec{g}(u, v), \cdots, \vec{g}(u + \Delta u, v + \Delta v)$. If Δu and Δv are very small this quadrilateral whose sides from the vertex $\vec{x} = \vec{g}(u, v)$ to the adjacent vertices are (almost) the vectors $\vec{g}(u + \Delta u, v) - \vec{g}(u, v) \sim \dfrac{\partial \vec{g}}{\partial u} \Delta u$ and $\vec{g}(v, v + \Delta v) - \vec{g}(u, v) \sim \dfrac{\partial \vec{g}}{\partial v} \Delta v$. These two vectors are (essentially) tangent to the surface S at $\vec{x}$; therefore, their cross product is a vector normal to S at $\vec{x}$, whose norm is the area of the parallelogram they span:

$$\Delta A = \left\| \frac{\partial \vec{g}}{\partial u} \times \frac{\partial \vec{g}}{\partial v} \right\| \Delta u \Delta v.$$

Thus, if R is a bounded subset of W, of the type we used when we defined two-dimensional integrals, we expect that

$$\text{Area of } \vec{g}(R) = \iint_R \left\| \frac{\partial \vec{g}}{\partial u} \times \frac{\partial \vec{g}}{\partial v} \right\| du dv. \tag{3.46}$$

We take (3.46) as the *definition of area* of such regions $\vec{g}(R)$ in a parametrized surface S. Observe that this equality has many features that are similar to the arc length we obtained in equality (3.17) of this chapter. We can, in fact, obtain (3.46) by considering polyhedral approximations to the surface (as we did with polygonal approximations to a curve). This treatment is rather more complicated than the one we used to obtain (3.17); thus, we decided to present the "reasonable" argument that led us to (3.46).

Let us develop a more explicit form of equality (3.46). We use the notation

$$\frac{\partial(\varphi, \psi)}{\partial(u, v)} = det \begin{pmatrix} \dfrac{\partial\varphi}{\partial u} & \dfrac{\partial\varphi}{\partial v} \\ \dfrac{\partial\psi}{\partial u} & \dfrac{\partial\psi}{\partial v} \end{pmatrix} = \frac{\partial\varphi}{\partial u}\frac{\partial\psi}{\partial v} - \frac{\partial\varphi}{\partial v}\frac{\partial\psi}{\partial u}$$

where $\varphi = \varphi(u, v)$ and $\psi = \psi(u, v)$ are functions of u and v. By (3.45) applied to $\vec{g}(u, v) = (x, y, z)$

$$\frac{\partial\vec{g}}{\partial u} \times \frac{\partial\vec{g}}{\partial v} = \begin{vmatrix} \vec{i} & \vec{j} & \vec{k} \\ \dfrac{\partial x}{\partial u} & \dfrac{\partial y}{\partial u} & \dfrac{\partial z}{\partial u} \\ \dfrac{\partial x}{\partial v} & \dfrac{\partial y}{\partial v} & \dfrac{\partial z}{\partial v} \end{vmatrix} = \frac{\partial(y, z)}{\partial(u, v)}\vec{i} + \frac{\partial(z, x)}{\partial(u, v)}\vec{j} + \frac{\partial(x, y)}{\partial(u, v)}\vec{k}.$$

Thus,

$$A(\vec{g}(R)) = \text{Area of } \vec{g}(R) = \iint_R \sqrt{\left|\frac{\partial(y, z)}{\partial(u, v)}\right|^2 + \left|\frac{\partial(z, x)}{\partial(u, v)}\right|^2 + \left|\frac{\partial(x, y)}{\partial(u, v)}\right|^2} \, du \, dv. \qquad (3.47)$$

This formula is more complicated than (3.17); however, it is clearly an extension of it sharing many of its features. As was the case for (3.17) we should verify that the formula (3.47) is independent of the parametrization. As was the case for curves, this follows easily from the change of variable formula for integrals over regions in $\mathbb{R}^2$. We announce the general n-dimensional result:

Theorem 3.41   Change of variables for multiple integrals.   *Let $\vec{g}$ be a $1{:}1$ map of class C^1 from an open set $U \subset \mathbb{R}^n$ onto an open set $V \subset \mathbb{R}^n$ whose derivative $D\vec{g}(u)$ is invertible for all $u \in U$. Suppose T is an open bounded set such that $\overline{T} \subset U$ and ∂T has content zero and, moreover, f is a real-valued continuous function on $\vec{g}(\overline{T})$, except for a set of content zero, then*

$$\int_{\vec{g}(T)} f = \int_T (f \cdot \vec{g})(\vec{u})|det D\vec{g}(\vec{u})|d\vec{u}. \qquad (3.48)$$

Authors' Biographies

PETER M. LUTHY

Peter M. Luthy got his Ph.D. from Cornell University; immediately following that, he was a postdoc of Guido's at Washington University in St. Louis from 2012 to 2015. He is now the chair of the math department at the College of Mount Saint Vincent, a small college in New York City. When he was still in St. Louis, Guido and he talked many times about a course he taught at WashU, year after year, for good freshmen. It was during this time we decided we should write a book based on the course he had designed, and Steve Xiao joined us in the project.

GUIDO L. WEISS

Professor Guido L. Weiss has had a very wide and interesting career. If one enters "Guido Weiss" in Google, several rather long articles about him will appear. We shall describe them in relatively few words that will give an idea of what he has done mathematically. His CV is very broad. All his university degrees are from the University of Chicago. After graduating with a Ph.D., he was on the faculty of DePaul University from 1956 to 1960. After this, from 1960 to 2015 his positions were all at Washington University in St. Louis. During his fifty-five years period he was very active as a mathematician and performed a considerable amount of research and occupied several academic positions. The list of publications (from 1955 to 2012) contains 121 papers (in one of the Google articles) and has published nineteen more articles and is now cooking two more.

His professional experience and honors are considerable. He received honorary professorships from the University of Milano (Bicocca), in Italy, the University of Barcelona, Spain, the Beijing Normal University, in China; the Chauvenet Prize and other similar prizes.

He has been invited to many lectureships all over the world. In particular he has travelled to China many times since 1980 and has "replaced" many Soviet mathematicians as a professor (nine of his forty-eight Ph.D. students are from China).

STEVEN S. XIAO

Steven S. Xiao earned a Ph.D. in Mathematics from the Institute of Mathematics, Chinese Academy of Sciences, Beijing, China. He once was one of Professor Weiss's post-docs and now takes care of all the computing in the department. He also teaches mathematics courses and computer programming courses in the university college.

Index

Printed in the United States
by Baker & Taylor Publisher Services